THE CELESTIAL COSMOS
ETHERS AND PLASMA

THE CELESTIAL COSMOS

ETHERS AND PLASMA

DR. DOUGLAS GABRIEL

Our Spirit, LLC
2025

OUR SPIRIT, LLC

P. O. Box 355
Northville, MI 48167

www.ourspirit.com
www.neoanthroposophy.com
www.gospelofsophia.com
www.eternalcurriculum.com

Library of Congress Control Number:
ISBN: 978-1-963709-17-9

Contents

Spiritual Science and Astronomy

It is incumbent upon Anthroposophists to be keenly aware of what is going on in the world and to meet it with the wit and wisdom provided through the study of Rudolf Steiner's teachings. Integrating Rudolf Steiner's cosmology with the scientific discoveries that happen every day in our modern world will reveal that Steiner was far ahead of his time. Steiner's clear expositions concerning the etheric formative forces still provide the best answer for an applicable Theory of Everything. His indications on the correct locations of the galactic and super-galactic centers of the Universe came decades before science confirmed his indications. In Steiner's *Astronomy Course* (GA 323), the center of mass of our solar system was indicated to not be in the center of our Sun, but actually is a point found by combining the mass of the Sun with the three inner planets as a counterpoint to the three outer planets. All three of these astronomical indications of Rudolf Steiner have been proven to be true by modern science along with his profound indications about plasma and zero point energy. There are hundreds of other examples of Steiner's 'predictions' that seemed crazy in his day but have now been found to be quite accurate and lead to new ideas about the forces and beings behind nature. Dr. Steiner's astrophysics indications point towards new energy sources and endless discoveries that will change the ways we look at the Cosmos and mankind's place therein.

An awake, integrated anthroposophist should be watching for and thrilled by the newest discoveries of science that align with Rudolf Steiner's indications given over a hundred years ago. One could say that Steiner was the unsung prophet of modern science, and a careful examination of his works will reveal future discoveries and revelations that will change our world and aid in the ascension of mankind. Steiner was 100 years ahead of his time, and we can use his foresight to help reimagine and redefine the world through a non-materialistic Spiritual Science that begins with the human being as the heart of a living Universe. Each one of us is, in fact, a copy of the Cosmos, a microcosm of the macrocosm. The human being should be the focus of all efforts of spiritual development, and those efforts should integrate with the modern world and scientific innovation, not avoid it.

Steiner gave a picture of the Essene community at the time of Jesus of Nazareth that showed the 'wrong way' to remove yourself from the outer world and live in a little paradise that makes sure Lucifer and Ahriman do not enter in. This seclusion and renunciation of the modern world made the Essenes prideful and caused them to reject Lucifer and Ahriman, subsequently keeping these forces in the outer world, far from the inner sanctum of the Essene elect. This type of 'sequestering of the Essenes' is often repeated by anthroposophists who hide-out in the Anthroposophical Society and believe that if they only associate with other anthroposophists they will not have to deal with Lucifer and Ahriman. This is sadly untrue and, in fact, causes Lucifer and

Ahriman to take hold of them even stronger when later they leave the inner group and interact with 'evil' in the real world.

Another tendency of anthroposophists is to develop the habit of talking at length about being a follower of Mani, who Steiner said will lead humanity in fighting evil in the future. This seemingly brave tendency of 'Manichean anthroposophists' is well intended but often misdirected. The challenge of our time is to 'face' evil, not 'conquer' evil. The time for conquering evil will not come for thousands of years. Even though *facing* evil in our time is the task of our age, especially the evil of egotism and scientific materialism. The ability to develop discernment concerning evil and good is one of our most important missions. This confusion causes many anthroposophists to retreat from the world after attempting to 'conquer' the evil of our times, often failing miserably. The grip of evil is overwhelming at this stage of evolution and only an initiate can cognize, absorb, and conquer evil due to its pervasive power over mankind. Many anthroposophists attempt to cognize evil but cannot understand its deep roots, and then fall back to the safe and secure arms of other anthroposophists. Like an Essene community during Jesus Christ's time, Waldorf communities and anthroposophic initiatives often try to ignore the evil world rolling by them as they eat their bio-dynamic vegetables and sing their Waldorf-style grace before dinner. Then, if they can have a little eurythmy after eating, they are 'right with the world.' Unfortunately, you can't hide from today's evil due to its all-encompassing pervasiveness.

Rudolf Steiner told us to read the *Kalevala*, the Finnish national epic, as a guidebook for the challenges of our time. Few anthroposophists have ever read this epic tale, but if they did, they would notice that the challenges presented to the three heroes are the same challenges we meet in our own age. It is recognition of evil and its control over your desires that is displayed in the plot of the *Kalevala*. The three heroes are fighting to win the Sampo, a free-energy device that produces gold, salt, and meal. The owner of this magic machine would be the wealthiest person on Earth. Each hero must personally struggle with their own 'double' before they can attempt an onslaught against evil to make, and later steal, the magic wonder-machine, the Sampo. I won't spoil the end of the story for you, but let's just say it is not such a happy quest, and the heroes are challenged to the limits of their being. One hero does eventually forge the magic Sampo, but it is taken by evil and eventually broken into many pieces. The story truly embodies the challenges of our age to face evil within and without, and to not *kill* evil, because it too is a part of your bodily constitution that you need to absorb and turn to good use. Just like Strader, in Rudolf Steiner's *Four Mystery Dramas,* develops a new machine with wondrous capacities. So too, each of us must find the physical and spiritual "engines" to power our lives making us rich in all things, like the owner of the Sampo or the Strader machine.

Evil in the world comes in many sizes and shapes but usually distills into an economic concern. Unless you live in a Camphill community or some other anthroposophic ideal setting that provides you with all you need to live, you must pay the devil his due—give unto Caesar what is Caesar's. Money, or evil 'Mammon,' should become a helpful tool instead of chains that bind your body, soul, and spirit. 'Earning a living' is a modern phrase that shows that there is little dignity afforded to the human being in the modern world. You don't deserve to 'have a living,' but must work hard to 'earn a living.' Who exactly are we earning that living from? Them—'the system'—the cold hard world—Ceasar—the Rat Race—the Devil—Ahriman, are all names given to the forces that demand that we 'earn a living.' You must pay the bills, but perhaps there is a wondrous machine that can help you 'create' the free energy that is all around us and ready for the taking.

What if you could fully integrate with the world and not have the effects of Lucifer and Ahriman hanging over your head telling you that you are not worthy of 'a living' and that you must compromise yourself and your integrity to pay the bills? What if understanding the forces developed through the study of Anthroposophy led you to a place where money was free, and simply 'living' your life provided all that you need? This pipe dream is possible for those who do not sacrifice their integrity to be 'removed' from the evil world of Lucifer and Ahriman. It is possible to use the forces of Lucifer and Ahriman for good and to make our lives better through Christ mitigating those polar-opposite forces.

If we follow Steiner's advice to fully enter the world and have sincere interest in all things going on in the world economically, politically, and spiritually, then we might be able to see the spirit unfolding and ride that 'wave' as a source of nourishment and wealth on all levels. For instance, there are new energy devices that have been created that can revolutionize the world and quell the evil control of petro-chemical corporations. Aligning oneself with these budding technologies could not only cut the chains of economic enslavement, but they could also make you rich. When you discover that there are cars and generators that run on tap water alone, then Big Oil warmongers no longer have control over you—you face-off with evil and you win a battle, but not the war. If you start to sell these new energy devices, you could benefit from a business that not only makes you money but also supports your beliefs in clean energy and reducing the pollution of gas-guzzling cars and radioactive power plants that use dirty fuels.

Water is life, and it can drive our world if we engender moral innovation and confront oil companies that foment 'oil-wars,' produce filthy fuels, and use the poisonous by-products to make pharmaceutical drugs. Having cognitive awareness of evil helps you understand the impending need to find new energy sources that could successfully allow you to consciously grapple with evil and eventually 'break the chains that bind' awareness of new Moral Inspirations that can take us beyond the enslavement of fossil fuels to a new day of natural energy that sets you and the world free. You can become the smithy who forges the new Sampo that provides all that is needed both in your own ascension body and in the new moral technologies that will help, not harm, human beings.

Some new energy devices are directly aligned with the spiritual scientific ideas of Rudolf Steiner. Many inventors credit the 'ethers' as the motive force behind their discoveries. Steiner's ideas on the ethers and plasma will inform technical innovation for centuries to come. He revealed the forces working in nature and from those indications have come a variety of inventions in many different areas of life. Just in medicine alone, Steiner and his followers have invented hundreds of new medicines that are very effective. A list of Steiner's discoveries would fill a book, and those indications will inform new technologies in the future. It is often the case that some new inventors are well-aware of Steiner's teachings and use them as inspiration for scientific innovation. From Steiner's ideas, we can develop a cosmology that informs modern science and the Science of the Spirit. The only difference between spiritual and material "science" is whether the human being stands as the foundation and touchpoint for scientific inquiry. The human being must be seen as the 'question' and the 'answer' for all modern problems of energy, innovation, and discovery. As the wise words of the Greek Mystery schools indicated, 'Oh man, know thyself and thou shalt know the gods and the world.'

Rudolf Steiner's Cosmos

Rudolf Steiner's Spiritual Science presents us with a comprehensive cosmology that is a Theory of Everything in accordance with spiritual laws and physical laws, while placing moral concern for mankind at the heart of all knowledge. Steiner's physical world is no 'Maya' of delusion to be ignored as evil, nor a dead entropic 'big bang' that is winding down to cold, entropic death. Anthroposophy is a new way to see beyond the 'veils' of nature to the beings and forces that create, maintain, and dissolve life in the Cosmos. Steiner informs us in *The Spiritual Beings in the Heavenly Bodies and in the Kingdoms of Nature* (April 1912, Helsingfors, GA 136): "The world of physical heavenly bodies represents the remains of the past deeds of the various beings of the hierarchies, and the aftereffects of their deeds extend into the present." Thus, when we look out into space, we are looking at the remains of the past left by spiritual beings and the insinuation of the future of evolution; a sort of time-capsule called Earth that is being pulled behind the Sun as it speeds towards the fixed star Vega—the solar apex.

We will hear in the Rudolf Steiner indications below that Spiritual Science revealed the true motion of the planets around (behind) the Sun. Dr. Steiner, in his *Light Course,* also revealed that light can move faster than the speed presently attributed to it and that light creates a continuum, inside of which it can propagate much faster than the frontal motion of light. The many insights Dr. Steiner has given concerning light, and the other ethers (warmth, sound, life, akasha, etc.) should make his 'revelations' something modern science would want to research. But it is only now that, 100 years later, many of Steiner's indications have been 'proven' by modern science. You will discover many of these 'prophecies,' and early discoveries of Steiner, in the quotations below. The application of these revelations is limited only by the Moral Imagination, Moral Inspiration, and Moral Intuition of the reader.

There is hope that human intelligence will win the battle against machine intelligence that is raging throughout the world. We believe it will. We also believe that divine intelligence, what Rudolf Steiner calls 'Intellectual Clairvoyance' or Moral Imagination, will lead human intelligence into the realms of Angels, the Spirit-Self of the human being. When this is accomplished, it will be astronomy, and its sister science of embryology, that will begin changing the materialistic direction of science. Dr. Steiner tells us in the lectures entitled, *Background to the Gospel of St. Mark* (GA 124) that: "The science of astronomy is that science which, sooner than all others, may be restored to spirituality."

Microcosm and Macrocosm

The Cosmic Origin of the Human Form, Rudolf Steiner, August 22, 1922, Oxford, GA 218

"The heart of a person, however, is in effect a Universe [Cosmos] turned inside out, and that is how the physical world is joined to the spiritual."

Etherization of the Blood, Rudolf Steiner, October 1, 1911, Basel, GA 130

"Electricity is light in the sub-material state. Light is there, compressed to the utmost degree. An inward quality too must be ascribed to light; light is itself at every point in space. Warmth will expand in the three dimensions of space. In light there is a fourth; it is of fourfold extension—it has the quality of inwardness as a fourth dimension."

The Fall of the Spirits of Darkness, Lecture XIV, Rudolf Steiner, Dornach, October 28, 1917, GA 177

"After the Mystery of Golgotha, we can speak of that which works in the soul as Christ-Impulse, as the spiritual sunlight, as though we saw a self-illumined body before us radiating its light from within."

Concerning the Astral World and Devachan, Rudolf Steiner, Lesson III, *The First, Second, and Third Logoi*, Summer, 1903, Berlin, GA 88

"All life in its manifold forms came forth from Unity, from the Logos. All diversity resides in it, but still unseparated, undifferentiated, hidden. Just as it becomes knowable, perceives itself as a self, it emerges from the Absolute, from the undifferentiated, and creates the not-self, its mirrored reflection, the Second Logos. It ensouls and enlivens this mirrored reflection; that is its third aspect, the Third Logos."

"Thus, the First Logos, the undifferentiated, in which life and form remain unseparated, is to be regarded as the Father. Time begins with his existence; he separates his mirrored reflection from himself, the form, the Feminine, which he then fills with his life, the Second Logos; and from this ensouling emerges the Third Logos as the Son, as enlivened form, as Father, Mother, and Son."

Universal Spirituality and Human Physicality, Bridging the Divide, Rudolf Steiner, Lecture II, *Morality as the Source of World-Creative Power*, December 1920, GA 202

"Our *thinking* is connected with the Cosmos, our *willpower* with the Earth, and our rhythmic life—the element of *feeling* that mediates between them—balances between the two, between the heavenly and the earthly. *Thinking* in ancient Mystery wisdom was conceived as being permeated by the principle of *Beauty*, *feeling* was regarded as being imbued with *Wisdom*, and *willpower* was thought to be penetrated by *Vigor*."

"Love is a will element, and fraternity has to pervade economic life. Freedom is the distinctive quality of waking life, whereas fraternity is that of sleep life."

"From our present time onwards, *beauty* imbibed from the universe becomes *Imagination* for humanity, while *vigor* becomes *Intuition* when grasped through free, independent human willpower, and *wisdom* becomes *Inspiration*."

***Universal Spirituality and Human Physicality, Bridging the Divide*, Rudolf Steiner, Lecture V, GA 202**

"We take possession of the light in a sense—the light that you otherwise perceive externally you also incorporate into you, differentiating it and working within it. This is your thinking faculty—activity within light. You are a being of light, but do not know it because you live in that element. Nevertheless, the thinking you develop is 'a life' within light, which you can certainly perceive once you regard it from without. *The universe is inwardly light; and observed from without, thought.* The human head is inwardly thought, and seen from without, light. So, you can see that *light and thought belong together*. They are in fact the same thing, just seen from different perspectives. The past is what gleams forth in the beauty of light—though naturally *light also signifies here everything that manifests, including tone and warmth*. The nature of light is thought, and that of darkness is will."

***Universal Spirituality and Human Physicality, Bridging the Divide*, Rudolf Steiner, Lecture VI, December 1920, Dornach, GA 202**

"What lives around us in nature is the outcome of former moral occurrences, in the form of *dying thought worlds manifesting as light*. But in their most intrinsic nature our thoughts are the residue of an ancient past but permeated by willpower. Pure thinking is pervaded by willpower. In the light, lives dying world thoughts, whereas, in the forces of weight live worlds growing from germinal willpower."

***Initiation Science and the Development of the Human Mind*, Rudolf Steiner, Lecture II, GA 228**

"Everything that happens in the universe is reflected by the Moon as if in a mirror. The things that happen inside the Moon remain hidden. The great, most ancient wisdom of which people speak is essentially a legacy from these Moon spirits who today live hidden lives in the Moon. They have withdrawn there."

***The Sun Mystery and the Mystery of Death and Resurrection*, Rudolf Steiner, Lecture IV, *Historical Changes in the Experience of Breathing*, March 26, 1922, Dornach, GA 211**

"Some individuals remained slightly aware that cosmic spirit and soul had once entered human beings through their breathing. These individuals gave the name *sophia* to the results of the joint activity of sense perception and breathing. Breathing itself, however, was no longer perceived. Its spiritual content was deadened—"paralyzed" would be a better way to put it—by sense perception. The Greeks perceived activity in the head as I have just described it, and they called it *sophia*. Those with a special fondness for developing *sophia* in themselves called themselves philosophers. To them, the word "philosophy" expressed the inner experience of living *sophia*."

"Thus, wisdom and faith flowed together in the human being. Wisdom streamed toward the head, and faith lived in the entire body. Wisdom was the content, or ideas; faith was the power of this content. The two belonged together. *Sophia* or Wisdom is diluted inhalation; *pistis* or Faith is condensed exhalation. In later times, Wisdom continued to be diluted until it became science. Similarly, inner strength continued to condense until people felt only their bodies and lost the awareness of what Faith, or *pistis*, actually is. Because they no longer sensed the connection between *sophia* and *pistis*, people began to separate understanding based on outer sense perceptions from subjective, internal belief. First there was *sophia*, then *scientia*, or ordinary science, a diluted form of *sophia*. We might also say Sophia was originally an actual spiritual being that humans experienced as inhabiting their heads. In ancient times, it was appropriate for people to become intoxicated through breathing, but it is not appropriate for modern people to become intoxicated with words. Words must pervade us with wisdom, life *sophia*."

***Planetary Spheres and their Influence on Man's Life on Earth and in the Spiritual Worlds*, Rudolf Steiner, Lecture I, *The Cosmic Origin of the Human Form*, August 22, 1922, Oxford, GA 211**

"So does man build his entire body both from the movements of the stars and from the constellations of the stars in the great Universe. *The human body is indeed an image of the world of stars.* Much of the work we have to do between death and new birth consists in the building of our own body from the Universe. *Man, as he stands on Earth is indeed a shrunken Universe.* Science is so naive as to suppose that the human form is produced from the physical germ-cell alone. Suppose a man is looking at a magnet-needle, one end of which always points to the North and the other to the South. Perhaps another man to whom this is explained does not believe it but begins looking for the cause inside the magnet-needle only, failing to see that the whole Earth is acting as a magnet. It is no less naive when someone thinks that man originates from the physical human germ-cell, whereas in fact he springs from the entire Universe.

Moreover, his life of soul and spirit between death and rebirth consists in working with the spiritual Beings—working at the super-sensible form of man, which is created first in the ethereal and astral realm and only then shrinks and contracts till it is able to be clothed in physical material. Man, in reality, is but the scene of action of what the Universe, and he himself with his transmuted powers, do thus achieve when the physical body in its true nature is being formed."

"But if one could do this, *taking hold of him in the inmost heart and turning him inside out like a glove, then man would not remain man as we see him here; he would enlarge into a Universe.* For if we have the faculty to concentrate in a single point within the heart and thence to turn ourselves inside out in spirit, we simply do become the Universe which in the normal course we experience between death and a new birth. Such is the secret of the inner man. It is only in the physical world that he cannot be turned inside out. The heart of man, however, is in effect a Universe turned inside out, and that is how the physical and earthly world is really joined to the spiritual."

The Gospel of St. John in Relation to the Other Three Gospels, Rudolf Steiner, Lecture XIII, _The Cosmic Significance of the Mystery of Golgotha_, July 6, 1909, Kassel, GA 112

"But every physical event has its spiritual counterpart, and as the blood flowed from the wounds on Golgotha there occurred a corresponding spiritual event: at this moment it happened for the first time that rays streamed forth from the Earth into cosmic space, where formerly there had been none. Darker and darker had the Earth become with the passing of time—up to the event on Golgotha. *Now the blood flows on Golgotha—and the Earth begins to radiate light.*"

"If in pre-Christian time some clairvoyant being had been able to observe the Earth from a distant cosmic body, he would have seen the Earth's aura gradually fading out, and darkest immediately preceding the event on Golgotha. Then, however, he would have seen it shine forth in new colors. The deed on Golgotha suffused the Earth with an astral light that will gradually become an etheric and then a physical light. Every being in the world continues to evolve. *What is today the Sun was first a planet; and just as the Old Saturn became a Sun, so our Earth, now a planet, will gradually develop into a sun.* The first impetus in this direction was given when the blood flowed from the wounds of our Redeemer on Golgotha. The Earth began to shine—for the time being astrally, visible only to the seer; but in the future the astral light will become physical light, and the Earth will be a luminous body, a sun-body."

"I have explained repeatedly that no new cosmic body comes into being through the agglomeration of physical matter, but through the creation, by a spiritual being, of a new spiritual center, a new sphere of activity. The formation of a cosmic body begins in spirit. *Every physical cosmic body was first spirit.* What our Earth will ultimately become consists at present of the astral aspect of its aura which began to ray forth from the Earth at the time we are here considering: that is the first nucleus of the future sun-earth. But what a man of that time would have perceived with his misleading senses is a phantom: that has no truth, it dissolves, it ceases to be; and *the farther the Earth moves toward its sun-state, the more will this Maya be consumed and perish in the fire of the Sun.* But through having been suffused at that time with a new force, through the newly created possibility for the Earth to become a sun, it became possible as well for this same force to permeate man. This was the first impulse toward what I described yesterday: the radiating of the Christ force into the etheric human body; and thanks to the streaming in of this astral force it could start absorbing new vitality such as it will need in the distant future."

"So if you will visualize the period in which the event of Golgotha occurred and then compare it with a later period—that is, if you compare a future condition of humanity with that which prevailed at the time of the event of Golgotha—you will find that at the time the Christ impulse intervened; the Earth, of itself, had nothing left to infuse into the etheric bodies of men. Some time later, however, the etheric bodies of those who had found a contact with the Christ-Impulse were irradiated: men who understood the Christ absorbed the radiant force that has been in the Earth ever since—the Earth's new radiance. They have taken the light of Christ into their etheric bodies. The Christ light streams into the etheric bodies of men."

"It is the possibility that was given at that time, as an effect of the Christ-light, for something new to appear, something living and breathing and immortal, something that can never perish in death. While men on Earth are still misled by the illusory image of death, this new factor will nevertheless be rescued from death, will have no part in it."

"Ever since that time, then, the human etheric body has held something that is not subject to death, to the death forces of the Earth. And this something which does not die with the rest, and which men gradually achieve through the influence of the Christ-impulse, now streams back again—out into cosmic space; and in proportion to its intensity in man it generates a certain force that flows out into cosmic space. And this force will in turn create a sphere around our earth that is in the process of becoming a sun: a sort of spiritual sphere is forming around the Earth, composed of the etheric bodies that have come alive. The Christ light radiates from the Earth, but there is also a kind of reflection of it that encircles the Earth. What is here reflected as the Christ-light, appearing as a consequence of the Christ event, this is what Christ called the Holy Spirit. Just as the event of Golgotha provided the first impetus for the Earth to become a sun, so it is true that beginning with this event the Earth began to be creative, surrounding itself with a spiritual ring which, in turn, will in the future develop into a sort of planet circling the Earth."

***Spirit as Sculpture of the Human Organism*, Rudolf Steiner, Lecture XV, *Human Experience of the Etheric Body*, December 7, 1922, Berlin, GA 218**

"The world within us that we perceive as full of spiritual beings becomes a kind of image or reflection in earthly existence. Here we see the sensory reflection of those beings whom we perceive within us between death and a new birth. We do not see the beings themselves as such but instead their dwelling places, and these are—since there are always a host of these beings together—the starry heavens around us. What therefore are we describing when we speak of the stars or the Sun, for instance, when we speak with full knowledge rather than the 'mole' vision of ordinary awareness we possess between birth and death? The Sun offers a certain image to our sensory gaze, but what we see as the image of the Sun here is something we perceive between death and new birth as a realm of spiritual beings. We see a realm of spiritual beings that corresponds to the Sun we see from an earthly perspective. And this is true too of the other stars. So, our spiritual awareness between death and a new birth is a cosmic one. There, we are not enclosed within our skin as we are here but have truly become the whole world. But we should not think of this in spatial terms. We become the whole universe and carry the starry heavens within us. Just as we have our lungs, heart, and stomach inside us here, so between death and a new birth we carry the Sun, the Moon, Saturn, and the other planets and stars inside us as our inner organs; but they are spiritual beings. We carry in us their spiritual correlation, their spiritual archetype.

"If we were always in this condition in the spiritual world, we would never come back to ourselves, would always feel ourselves to be one with the world of the higher hierarchies. But that cannot be. That would be like always only breathing in and never out. This is why our life between death and a new birth consists of a rhythmic alternation: a life within these higher hierarchies and—in a cosmic form of consciousness—in looking outwards, which there means coming back to ourselves. Just as we alternate here between inbreath and outbreath—or I could also say between being awake and being asleep—so there, we alternate between an experience of the hierarchies, the world of spirit, and an experience of ourselves, where we come to ourselves and feel the loneliness of our contracted soul. Thus, we have a recurring rhythmic alternation of experience between being spread out through the whole universe and coming back to ourselves."

"The physical human body standing before us may not immediately betray the fact that it is the result of the work of divine beings in collaboration with the human being during this existence between death and a new birth. But older civilizations were aware of this when they called the body a 'temple of the gods.' As little as we normally realize this with our ordinary consciousness here on earth, *this human body is the most complex thing existing anywhere in the universe.* A single human body is the confluence and totality of the work of countless beings, of whom we too are one. We help to work on the body with which we clothe ourselves in an earthly incarnation. We would not be able to elaborate it individually for ourselves but have to create it in community with countless spiritual beings of the most diverse ranks.

"From an earthly perspective, we are used to regarding a seed as something that is small to start with that unfolds to occupy greater physical space. If we call what we elaborate between death and a new birth the 'spirit seed' or 'spirit germ' of the physical body, this spirit seed is, in fact, as vast as the universe and grows ever smaller physically as it passes through the embryo stage. *The small fertilized human ovum contains a reflection of this great spirit seed that has been elaborated by the human being together with higher beings.* If we use spiritual vision to study the world human beings live in between death and a new birth, we find how the tasks of the macrocosm give rise to the microcosm, the human body, which is continually configured anew. And this task is more sublime than all cultural endeavors accomplished by human beings during life on Earth. The life we spend there, working from the universe upon the human seed, is more varied and richer than what we accomplish here on earth."

"Specifically at the culmination of our experience between death and a new birth, which I called 'the midnight hour of human existence' in one of my *Mystery Plays*, the human being experiences his inner life as the starry heavens he sees here on Earth in reflected form, the heaven of fixed stars. The world of fixed stars or their representative—as older cultures called it—the zodiac is, from our earthly perspective, the physical image or reflection of the spiritual world in which we live between death and a new birth, and which we then experience as our inner world."

The Realm of Sleep and the Cosmos

Spirit as Sculptor of the Human Form, Rudolf Steiner, Lecture I, *Experiences During Sleep. Their Spiritual Context and Significance for our Daily Life*, Oct. 9, 1922, Stuttgart, GA 218

"Please note that I am not saying we dwell within planetary motions during sleep or are connected directly with them, but rather that we dwell in a kind of reflection or replica—if you like, a miniature of our planetary cosmos or its movements. As the life of the soul inhabits blood and circulation during the day, so *at night our soul life inhabits a replica of the planetary movements of our solar system.* At night, by contrast, we have to say that in our life of soul a replica of the movements of Mercury, of Venus, and Jupiter circulate in us.

"Thus, from the moment we fall asleep to when we wake up, *our soul life is in a sense a small planetary cosmos.* From our personal, human existence during the day we pass into a cosmic existence at night. ...In order for us to embark on new soul life the following day, we need the impetus received by experiencing a reflection of the planetary cosmos during the night. On awakening, the after-effect of what we experienced in these reflections of planetary movement at night is implanted or instilled in us. And this is what connects the cosmos with our individual life. If this resonance of our nightly experiences was not present in us when we wake up in the morning, the powers we need could not stream into us in the right way to endow us with proper awareness."

Birth of the Zodiac

***The Influence of Spiritual Beings on the Human Being*, Rudolf Steiner, Lecture II, January 27, 1908, GA 102**

"In the lecture today, we shall make a rather far-reaching sweep into cosmic space. This will reveal to us, in broad outline, the inner course of world evolution, and at the same time its intimate connection with human evolution on the Earth. *Everything in the universe is interconnected.* To be able to follow these complicated connections naturally takes a long, long time, and it is only very gradually that man can find his way, so to speak, into the intricate workings of the cosmos.

"In previous lectures you have heard how certain beings who have their abode on other cosmic bodies exercise an influence upon our own life, how they are related to what we call lymph, to the digestive fluids, also to our sense-perceptions. This will have given you a picture of the wide-spread operations of the spirit throughout cosmic space. We shall study a different aspect of these things today, reminding ourselves, to begin with, that our Earth, like man himself, has passed through different embodiments and will pass through others in times to come.

"We look back to three previous embodiments of our Earth: to the immediately preceding embodiment which we call the Old Moon (not to be confused with our present Moon); then to that of the "Old Sun"; and still further back to that of "Old Saturn." And looking forward we see prophetically that our Earth will be transformed into a "Future Jupiter," a "Future Venus" and a "Future Vulcan."

"These are the successive embodiments of our planet Earth. If you give a little thought to these stages of our Earth's evolution, you will realize that what in occult science we call a "sun" is—like our present Sun—a heavenly body around which a number of planets revolve. When, apart from this, we also speak of a planetary sun-existence, saying that our Earth itself, in an earlier state of evolution, was "sun," we imply, in a certain respect, that the Sun which is today the center of our planetary system, was not always a sun. It has advanced, so to speak, to the rank and dignity of a sun in the Cosmos. It was once united with the substances and forces contained in our Earth and then, taking away, as it were, what was the best and most capable of the highest development, it separated from the Earth, leaving us, together with certain forces which were destined for a slower evolution, behind. The Sun took with it certain higher beings and together with these higher beings established itself at the center of our system. Therefore, two stages earlier, what is contained in the Sun today had a planetary existence only and it has risen from this to the form of existence belonging to the fixed stars. This will show you what mighty changes in evolution take place in the universe. At the outset, a sun is not a sun. *A fixed star has not, from the very beginning, been a fixed star, but has had to pass through the lower school of planetary existence.*

"Now you may quite naturally ask me: What, then, happens when a fixed star evolves to a further stage? As truly as the sun-existence—a fixed star existence—has risen from a planetary existence, so truly does its evolution proceed to further stages of life in the Cosmos. We shall of course understand this evolution still better if we study the further evolution of our Earth.

"It is true that for a certain period of its cosmic evolution our Earth has been separated from the Sun. The Sun and its beings advance along a more rapid evolutionary path. Our Earth and the beings belonging to it take a different course. But these beings, and the Earth as a whole, will one day have progressed to the stage where union is again possible with the Sun—after a separate existence has enabled them to complete and perfect their present phase of development. For our Earth will again unite with the Sun. *During the stage of Earth-existence itself, the Earth will reunite with the Sun, just as during the same phase of evolution it separated from the Sun.* But during the [Future] Jupiter stage there must again be a separation. The Earth-beings must again be separated from the Sun during the [Future] Jupiter-condition. Again, there will be a reunion, and during the [Future] Venus-condition our Earth will be united permanently with the Sun, will have been taken up for all time into the Sun. During the [Future] Vulcan-condition, our Earth will itself have become a sun within the Sun and have contributed something to the sun-evolution, will have added something which, in spite of their higher rank, those beings who have always remained in the Sun, could never themselves have achieved.

"Earth-existence was necessary in order that men might evolve as they have evolved, with a consciousness that alternates between waking and sleeping. This is connected with the separation from the Sun. Beings who live always in the Sun do not have day and night. The sense-consciousness which we call the "clear consciousness of day", and which in times to come, will evolve into higher conditions, carries with it into the sun-evolution the fruits of experiences connected with the things of outer physical space. In this way, the Earth-beings give something to the Sun, enrich the Sun. And out of what is thus acquired on the Earth, augmented by what is acquired on the Sun, the [Future] Vulcan-existence comes into being. This [Future] Vulcan-existence is actually a higher condition than that of our present sun-existence. The Earth evolves, the Sun evolves, until they can unite to constitute the [Future] Vulcan-existence.

"You may ask me: When a planet has evolved in this way to a sun-existence, what does this sun become in the course of further cosmic evolution? When our Earth reaches the

[Future] Venus-condition, it will itself have become Sun, and all the beings on [Future] Venus are sun-beings—actually at a higher stage than the beings of the present Sun. What, then, is the further stage of such planetary evolution?

"The following will seem grotesque, even preposterous, to those whose concepts are rooted in modern astronomy. Nevertheless, it is a truth of cosmic evolution that when a planet like our Earth has risen to sun-existence, when it has gradually achieved union with the Sun and even sun-existence is transcended, there arises, as a still higher stage of evolution, something that in a certain sense you can perceive in the heavens: there arises what we today call a "Zodiac"—it is the stage higher than that of a fixed star. Thus, when beings are no longer restricted to the form of existence belonging to a fixed star but have expanded their evolution so powerfully that it extends beyond fixed stars, and the fixed stars lie like bodies embedded in it—then a higher stage is reached, the stage of Zodiac-existence. The forces which work from a Zodiac upon a planetary system themselves evolved, in former ages, into a planetary system and have advanced to the stage of a Zodiac.

"And now cast your minds back to the Old Saturn evolution, the first embodiment of our Earth. This Old Saturn once glimmered, as it were, in cosmic space, as the first herald of the dawn of our planetary existence. You know, too, that on this Old Saturn the first germinal inception of our physical body was brought into being. Even at its greatest density this Saturn was not nearly as physically dense as our Earth. It was a condition of utmost rarefication. That which today permeates all beings as warmth—known in occultism as "fire"—was the matter of Old Saturn. We may picture to ourselves that around this Old Saturn, this first, dawn-condition of our planetary system, there were the constellations of the Zodiac—but not yet as they are today.

The single stars composing the Zodiacal constellations around Old Saturn were scarcely to be distinguished from each other. They glittered only very faintly, like beams of light streaming out from Old Saturn. The best way to picture this is to think of Old Saturn encircled by beams of light, just as our Earth is encircled by a Zodiac. And in the course of Earth-evolution itself, these light-masses developed into the present star clusters comprised in the Zodiac. So that *the Zodiac*—to use an abstract expression—*has differentiated out of that original ocean of flame.* And from what did this ocean of flame itself arise?

"It arose from the planetary system [Manvantara] which preceded our own. Old Saturn itself was preceded by planetary evolutions in an age which, speaking in the sense of occult astronomy, can by no means be described as "time" as we understand time, for its character was rather different. But for the human mind today the concept is so fabulous that we have no words with which to express it. Speaking in analogy, however, we can say that the forces which preceded our planetary system in an earlier cycle of planetary existence went forth in the light-streams, and out of a small portion of matter gradually gathering together at the center, this first, dawn-condition of the Earth arose; this was Old Saturn and the forces contained in the Zodiac radiated down upon it from the cosmic All.

"Something rather remarkable comes to light when we compare planetary existence with zodiacal existence. The occultist makes use of two words to indicate the difference between them. He says: *Everything that is contained in the Zodiac is under the sign of "Duration"; everything that is comprised within planetary existence is under the sign of "Time."* You can get an idea of what this means if you remember that not even the farthest reaches of the mind can conceive of changes having taken place in the Zodiac.

Each single planet may have undergone considerable change through long and greatly differing periods of evolution; the forces working in the Zodiac remain, relatively speaking, fixed and permanent. These concepts can, in any case, be only relative. The only difference in these changes which we can conceive is in respect of the speed. *Changes in the Zodiac take place slowly; changes in the planetary world and even in the existence of a fixed star take place rapidly—in comparison*, that is to say, with what happens in the Zodiac.—The difference is always relative, only relative. As far as human thinking is concerned, we can say that planetary existence belongs to the sphere of the Finite, whereas Zodiacal existence belongs to the sphere of Infinitude. This, as already said, must be taken in the relative sense, but for the present it is sufficiently accurate.

"And now I would ask you to pay special attention to the following: What has been achieved in a planetary existence and has become sun, ascends to "heavenly" existence, becomes zodiacal existence. And having reached zodiacal existence, what does it do? It offers itself in sacrifice! Please take account of this particular word. The first dawn-condition of the Earth, Old Saturn, arose in a mysterious way as the result of sacrifice on the part of the Zodiac. *The forces which caused the first, rarefied Old Saturn-masses to gather together were those which streamed down from the Zodiac, producing on Saturn the first germinal inception of physical man.* This continued without cessation. You must not picture it as happening only once. Fundamentally speaking, what is happening continuously is that within what we call a planetary system, the forces which evolved to a higher stage, after having themselves passed through a planetary system, are sacrificed. We can say in effect: *what is at first contained in a planetary system evolves to a "sun" existence, then to zodiacal existence and then has the power to be itself creative, to offer itself in sacrifice within a planetary existence.* The forces from the Zodiac "rain" down continuously into the planetary existence and continuously ascend again; for that which at one time became our Zodiac must gradually ascend again. The distribution of forces in our earth-existence may be conceived as follows: on the one side forces are descending from the Zodiac and, on the other, forces are ascending to the Zodiac. Such is the mysterious interplay between the Zodiac and our Earth. Forces descend and forces ascend. This is the mysterious "heavenly ladder" upon which forces are descending and ascending."

Spirit as Sculptor of the Human Body, Rudolf Steiner, Lecture VIII, _The Battle for Human Nature Between Luciferic and Ahrimanic Beings_, November 16, 1922, London, GA 218

"In the other direction are the beings who incorporate themselves in air and warmth [fire]. These also, like the enemies of Jahve, are immature. And the corresponding mature beings are in their case beings who dwell on Mars, Jupiter, and Saturn. And so, we find these air-fire beings making sallies from their strongholds not only upon the ahrimanic powers, but upon the influences that should be continually reaching man from Mars, Jupiter, and Saturn.

"Now, the influences of these more distant planets,—or rather, of their spiritual beings—are to be found particularly in the eyes, in the ears,—in short, in the sense organs of man. So that, whilst Moon and Venus exercise their influence in the interior organs of man's body, Saturn, Jupiter, and Mars work on man's exterior, work in his sense organs. The influences, for example, of Saturn are to be met primarily in the human eye.

"These beings,—Saturn beings, Jupiter beings, and Mars beings—have it as their special concern to make man a real Earth man; that is to say, they want first to give him senses that are properly inserted into the human organism and that remain at its surface, and then to supply him with nerves that run from the senses and extend inwards into the organism. *Saturn gives the senses, Jupiter gives their continuation in the nerves, and Mars exerts the kind of control that endows man, for example, with the faculty of speech.* The whole aim and purpose of these beings is to furnish man with all that is on the surface of his body. For the senses, and the nerves too, have come about through a 'turning outside in' of the human skin.

"Jupiter, Saturn, and Mars are however resisted in their activities by the air-fire beings of whom we were speaking. Here again, a furious fight goes on all the time. *The air-fire [warmth-light ether] beings sit fast, so to speak, in their airy strongholds and display their power and might in the fiery flashes of lightning [plasma].* They would like to make the whole of man what he should properly be on the surface only, so that the entire physical being of man should partake of the nature that is actually assigned only to eye and ear and nose. They would like to pour the surface of man's organism right through him, to make him all surface, so that he would do nothing but see and hear, never eat nor drink, but only see and hear, be in fact a kind of angel being.

"The Mars, Jupiter, and Saturn beings, who work as we have seen in man's senses, acquit themselves most praiseworthily—if I may employ the expression in speaking of such sublime beings—in the world of external nature. For they permeate what to our eyes appears mere Nature, with morality. In this manner they bring morality to man; for it is actually so, *morality enters into us through the senses.* When therefore the air-fire beings seek to permeate man through and through with his sense nature, it is with the intention that man, seeing nothing but what is moral, may become a moral automaton.

"If we look out on the world of Nature, we can know that *whatever manifests as forces in that world comes from the Mars beings, whatever manifests as natural law [thought] from the Jupiter beings, and whatever manifests as color and sound from the Saturn beings.* And the air-fire beings would have man become nothing but force, law (thought), color, and sound. They want man not to have a physical body at all, but to be insubstantial, rarefied; they would like him to be, as we said, an angel being.

"And so, you see, whilst in external nature Moon, Mercury, Venus, Jupiter, Mars, and Saturn live at peace with one another and are held in balance by the Sun, they wage a double fight for the possession of man. First of all, there is the conflict that goes on between the ahrimanic and the luciferic beings; and then we have on the one hand, the fight that is put up by the luciferic beings against the planetary forces beyond the Sun,—the Mars, Jupiter, and Saturn influences—whilst on the other hand the ahrimanic forces are waging war on the influences that proceed from Moon, Venus, and Mercury."

The True Motion of the Earth Around the Sun

Nature and Spirit Beings, Their Effects in Our Visible World, Part II, Rudolf Steiner, Lecture X, *The Relationship between Worlds and Beings,* April 29, 1908, Munich, GA 98

"You know, for example, that the planets, the celestial bodies, perform certain movements around themselves and around others. Why does this happen? The movement of the Earth around its axis was not always there. Why did it come about? The reason is that the human being, at its present stage of development, needs the alternation between day and night, between sleeping and waking. The macrocosm is most intimately connected to the microcosm—through the division of time, life is regulated. During the time of the Old Moon, it was quite different. There was a completely different time allocation, quite a different alternation between day and night, because the Old Moon moved entirely differently. Those beings, who nowadays direct these movements, have prepared these already in their own lives. Spiritual beings are behind those movements which are their deeds. In the future, mankind will recognize a deep wisdom in those movements."

"A deep wisdom lies in the so-called "orbit" of the Earth around the Sun. Man will one day realize that something incredibly significant is happening there. Aren't you surprised that I am saying "so-called" orbit. What is today taught in schools about the way the Earth moves around the Sun is only the result of a mathematical example. It is not absolutely true. One day, this explanation will also take on quite different forms. Even from a historical perspective, people could inform themselves that it is not like that. It is quite a strange issue with the system of Copernicus. He founded his beliefs on three basic principles, of which only two were adopted by today's science, and the third one was dropped under the table. In reality, the Sun races with high speed through space towards the constellation of Hercules. A movement, as it is usually described, that is only feigned by the fact that the planets also move along. *The true path of the Earth forms a corkscrew.* What is called the obliquity of the ecliptic is the gravity line between Sun and Earth. One has forgotten that the Earth turns once a year around the axis of the ecliptic, and this rotation combines itself with the corkscrew turn. Copernicus still differentiated between those two things, but nowadays it is not done anymore. The movement with the ecliptic was dropped. Therefore, it is not compatible with the facts when one says that the Earth turns around the Sun. In reality, it is a corkscrew movement."

"If the corkscrew line were a straight line, then progress would be immensely faster—Earth would have to travel along its path with incredible speed, and that would be exactly what the human being couldn't cope with."

"If the Earth would really pass through those spaces in a straight line, then man would become old immediately. As it is, the movement was amended in a wise way by the leading Spirits. The absolute progress is being delayed through a different way of movement. As you can see, there lies deep wisdom in the Cosmos—this Wisdom is the expression of the leading Spirits. We have been given angels and archangels as governors of our evolution. The forces that are working from incarnation to incarnation, that push man further, so that he will not become mummified, these are the governors of the future rotation times of [Future] Jupiter. Such spirits, that are standing above the human being and regulate his life, are therefore called the "Spirits of the Rotation of Time", because their deeds will find expression later in the rotation times of the celestial bodies. The way the stars move today can be seen as the result of what higher beings have done in former times. In today's humanity you can already recognize the future times of rotation. With this, a tremendous spiritual life enters celestial space, when we learn to look at it in this way."

"Today, we only wanted to look at the characteristics of beings up to the [Spirts of] Powers. We can imagine how the external world is the expression of something internal. When what is said here fills humanity again, then much will change. We have now an immense low in academic education. The speed of external progress is incompatible with spiritual life. This would lead to a tremendous low if such truths were not made known and used to enlighten science. People no longer know where to go with their materialistic science. Recently, a psychology book was published—one should not think that such a book has no effect just because the author is still unknown. The book explains that the law of the conservation of forces also applies to the soul, and that the inner manifestations of the soul consist only of a transformation of food. He roughly says, "For ten years, it has been known with certainty that the so-called law of the conservation of forces is identical with the effects of the nervous system. One can prove that all that man absorbs in the form of energy from the consumption of food, is completely identical to that which he produces in work. Since one can prove exactly the same thing happens inside man as elsewhere in the world, there can, therefore, be no soul being. We are only dealing with the conversion of food into energy, which is given off again to the outside."

"This is a very smart conclusion. Just as well one could say: In front of a Bank are two people counting money that is carried in and out. The amounts are equal, therefore inside the bank there are no staff. But is not staff needed to manage everything? This example is on the same level as the psychologist's opinion and a large part of what figures as science today. Everyone who even somehow considers this matter can imagine what a spiritual culture that thinks so little would lead to. It is necessary to possess spiritual knowledge, for it is only here that the real impulse is given to the development of humanity. If human beings do not find out what is behind the manifestations, then the world cannot be understood. One must arrive at the great, far-reaching, all-embracing laws, at the relationships of beings and worlds."

Man: Hieroglyph of the Universe, Rudolf Steiner, April 9—May 16, 1920, GA 201

Lecture I

"*Man can find absolutely nothing in the Universe unless he finds it first in himself.* He does not experience, for instance, the assuming of the upright position, being in the vertical; and so, he is not aware that he is in a vertical position for the simple reason that he moves together with the Earth in a certain direction which adheres to the vertical. Neither does he know that he makes his breathing movements, his digestive and eating movements, as well as other movements, in a direction through which the Earth also moves in a certain line. All this adherence to certain directions of movement implies an adaptation, a fitting into, the movements of the Universe."

"*The ability to think of the Universe as an organism depends on our learning to understand the hieroglyph of the organism we have before us.* We must learn to perceive Man as a hieroglyph of the Universe, for he gives us the opportunity of seeing near at hand how different are above and below, left and right, before and behind. We must learn this first in Man, and we shall then find it in the Universe."

Lecture II

"At the present period in human thought we compress the whole world within abstract lines of space, standing perpendicular to one another and forming the three dimensions of space, whereas in its life aspect this three-dimensional world proves to be much more complicated and much more concrete."

"After yesterday's suggestions, Dr. Stein has taken the trouble to construct a model showing the movements which result when we follow Man together with the Earth, or in

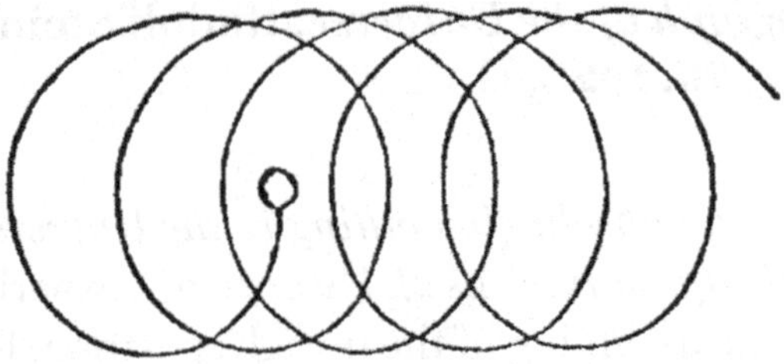

other words the movement of the Earth taken in its absolute sense. If instead of following this time the motion of plant-forces in spirals, I follow the movements described by Man with the Earth, I again come upon a spiral, but one which is progressive. This spiral gives us an illustration of the real movement of the Earth, and at the same time a picture of that of the Sun. Suppose for a moment that the Earth is here and the Sun there. An observer sees the Sun in this direction. (see diagram) The Earth progresses, but exactly in a line behind the Sun. When the Earth is here, the observer now sees the Sun in another direction. The Sun advances still further, the Earth follows, and once again the observer sees the Sun in the other direction. That is to say, he sees the Sun at one time on the right, and another time on the left, owing to the way in which the Earth follows the Sun."

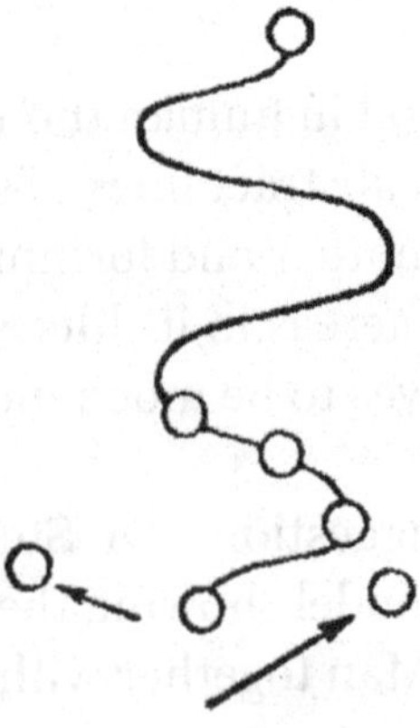

"This has been interpreted as demonstrating that the Sun stands still, and the Earth revolves round it. In reality, it is not so; the Earth moves along behind the Sun. The observer sees the Sun to the right when the latter has arrived at one point of the spiral path, while the Earth is here. Next, he sees the Sun to the left, then again right, then left, and so on. All this gives the observer, who judges by outward appearances and loses sight of his own movement, the impression that the Earth revolves round the Sun."

"From this you will realize how great a possibility of deception arises when one judges by exterior appearances; for here indeed a relativity of motion exists. We can really affirm that those who now calculate the apparent motion of the Sun do not perceive their own motion and omit to take into consideration the relation between the Sun and the Earth."

"I should like you to try to form a true idea of what I have said about course or motion in a screw-like line, because one must visualize, in a model such as this, the fact of the Earth following in the wake of the Sun; and then we shall be able to go on to what I should like us to attain tomorrow, namely a true understanding of the facts before us."

"But I will show you first how to gain a conception of the true relation of the Earth to the Sun—that the Earth actually follows the Sun in its path—by searching for the one thing that will show us this relationship, namely certain processes in the human organism connected with the representative of the Sun in man—the human heart. For it is by taking our start from the knowledge of Man that we must seek to attain to a knowledge of the Universe."

Lecture III

"If we thus seek in the world outside what we experience inwardly, we come indeed to the feeling that we are not really within our self at all, but that with our real Ego we are in the Universe, poured out into the Universe. Instead of searching behind the external Universe for 'vibrations', the atomists should seek their own Ego behind the phenomena and then try to find out how their own Ego is placed into the outer Universe is, as it were, poured out into it. Just as with color, we should try to ascertain whether we feel we must plunge into it or whether we feel ourselves repelled by it, so, as regards the structure of our organism, we should feel how the three directions, above and below, forwards and backwards, right and left, differ concretely from one another; we should feel how differently we experience them inwardly, when we project ourselves into the Universe. When we are aware of ourselves as Man standing on the Earth, surrounded by the planets and fixed stars, we begin to feel ourselves as part of all these; it is not a matter merely of drawing three dimensions at right angles, but of thinking concretely about the Cosmos and penetrating into the concrete reality of the dimensions."

"We distinguish in some way what takes place in the four realms,—what is above and below the plane of the Zodiac (Willpower), and what is right and left of the plane (Feeling); or again, we can consider what lies on this or on the other side of the plane (Thinking). We feel that something is connected with this differentiation, something of cosmic happening, namely, that which manifests in recapitulation, as we have it for instance in what we designate as the "course of the year."

"Man was at one time, as regards his inner conditions, bound to the outer course of the Universe, but he has broken away from it. Civilized Man today has almost entirely broken away from the course of outer Nature."

"We only learn the significance of certain connections which we read from the stars, when we understand the corresponding processes in our organism; for what lies within our skin is no other than a reflection of the organism of the outer world."

"Look at the principal course of the blood-vessels in the human organism. Seen from above it is like a looped line. Instead of drawing it, *we should follow the hieroglyphs inscribed in our own selves; for then we would learn to understand the nature of the qualities in the Universe outside.*

The heart does not work like a pump driving blood through the body, but that the heart is moved by the circulation, which is itself a living thing, and the circulation is in its turn conditioned by the organs. The heart, as can be followed in embryology, is really nothing more than a product of the blood's circulation."

"Just as the movement of the heart is the product of the life-force of the circulation, so the Sun is no other than the product of the whole Planetary System. The Sun is the result, not the point of departure. *The living co-operation of the solar system produces in the center a hollow, which reflects as a mirror. That is the Sun!* I have often said that the physicist would be greatly astonished if he could travel to the Sun and find there nothing of what he now imagines, but simply a hollow space; nay, even a hollow space of suction which annihilates everything within it. A space indeed, that is *less than hollow*. A hollow space merely receives what is put into it; but *the Sun is a hollow space of such a nature that anything brought to it is immediately absorbed and disappears.* There, in the Sun, is not only nothing, but less than nothing. What shines to us in the light is the reflection of what first comes in from Cosmic space—just as *the movement of the heart is, as it were, what is arrested there in the co-operation of the organs*, in the blood-movement, through the activity of thirst and hunger and so forth."

In order to make a few of the features of this cosmogony clear, we can draw a spiral; but this does not yet show the actual state of things. For to describe a few more features, we must make the spiral itself move spirally; we must make the line itself curve."

"*One must learn to regard the present conception of the world as an absolute delusion.*"

Lecture IV

"The fundamental nature and construction of the Universe cannot be conceived in its reality without continual reference to Man. Again and again, we must try to find in the Universe outside, what exists in one way or another in Man."

"I have already often drawn your attention to this correspondence between Man and the world in which he evolves. I have pointed to the fact that Man, in his rhythm of breathing—18 per minute—manifests something that is in remarkable accord with other processes of the Universe. We make 18 respirations per minute, which gives when calculated for the day, 25,920 respirations. And we arrive at the same number when we calculate how many days are contained in a normal life term of 72 years. That also gives about 25,920 days; so that *something may be said to exhale our astral body and Ego, on falling asleep and inhale them again upon waking*—always in conformity with the same number rhythm."

"Take the 18 respirations per minute, making 1,080 per hour and in 24 hours 25,920 respirations; that is, we must multiply: 18 X 60 X 24 in order to arrive at 25,920."

"And again, when we consider how the Sun moves—whether apparently or in reality, does not matter—advancing a little each year in what we call the Precession of the Equinoxes, when we consider the number of years it takes the Sun to make this journey round the whole Zodiac, once more we get 25,920 years—the Platonic Year."

"...We can in actual reality observe the 'breathing' of the Macrocosm. We only need to notice the path of the Moon in 18 years or, in other words, the Nutation of the Earth's axis [Chandler wobble]. The Earth dances, and she dances in such a manner as to describe a cone, a double cone, in 18 years, and this dancing is a reflection of the macrocosmic breathing. This takes place just as many times in the macrocosmic year as the 18 human respirations during the microcosmic day of 24 hours. So, we really have one macrocosmic respiration per minute in this Nutation movement. In other words, we look into this breathing of the Macrocosm through this Nutation movement of the Moon, and we have before us what corresponds to respiration in man."

"We have therefore not only to consider a world which is spread out before us, but another world which projects into, and permeates our own."

"Just as we have before us a second part of the human organism, when observing the breathing process, namely the rhythmic man, as opposed to the perceptive or head man, so we have in what appears as the yearly Moon motion, or rather the 18-year motion of the Moon [node], the identity between one year and one human respiration; we have this second world interpenetrating our own."

"The Sun is not a ball of gas; but in that place where the Sun is, there is something less than empty space—a sucking, absorbing body, in fact, while all around it is that which exerts pressure. Consequently, in *that which comes to us from the Sun we have not to do with anything constituting a product of combustion in the Sun; but all that has been transmitted to the Sun from the Universe is rayed back.*"

"*Where the Sun is, is emptier than empty space. This can be said of all parts of the Universe where we find Ether.*"

"In the same way, not only does matter become empty space, but it becomes negative, less than nothing—emptier than emptiness; it assumes a 'sucking' nature. Ether is sucking, absorbing. *Matter presses. Ether absorbs. The Sun is an absorbing, sucking ball,* and wherever Ether is present we have this absorbent force."

"Here we step over into the other side, the other aspect of three-dimensional space—we pass from pressure to suction. That which immediately surrounds us in this world, that of which we are constituted as physical man and etheric man, is both pressing and sucking or absorbing. We are a combination of both, whereas the Sun possesses the power of suction only, being nothing but ether, nothing but suction. *It is the undulating wave of pressure and suction, ponderable matter and ether, that forms in its alternation a living organization.* And the living organism continually breathes in the astral; the breathing expresses itself through the Moon's motion or nutation. *And here we begin to divine a second member or principle of the world's construction; the one member—pressure and suction, physical and etheric; the other, the second—astral. The astral is neither physical nor etheric but is continually inhaled and exhaled*; and the nutation demonstrates this process."

"Briefly, the facts are that if we notice the path of a fixed star and notice the point where the Sun stands over it, we find that at the end of 72 years the star occupies the same position on the 30th of December, while the Sun only reaches that point again on the 31st December. The Sun has lost a day. After a lapse of 25,920 years, this loss is so great that the Sun has described a complete revolution [around the zodiac] and once again is back upon the place we noted. We see therefore that in 72 years the Sun is one day behind the fixed stars. Now these 72 years are approximately the normal life period of Man, and they are composed of 25,920 days."

"Thus, when we multiply 72 years by 365, and consider the human span of life as one day, we have the human life as one day of the Macrocosm. Man is breathed out, as it were, from the Macrocosm; his life is one day in the macrocosmic year."

"...This revolution, this circle described by the Precession of the Equinoxes, indicating the macrocosmic year, as already known to the Egyptians thousands of years ago (for they looked upon this period of 72 years as very important), this apparent revolution of the Vernal point is connected with the life and death of Man in the Universe—with the life and death, that is, of the Macrocosm."

"We have, thus, three worlds interpenetrating one another, inter-related; but we must not attempt simply to combine these worlds from the point of view of causality. Three worlds, a three-fold world, as Man is a three-fold being; one, the world of sense surrounding us, the world we perceive; a second world whose presence is indicated by the motions of the Moon; and a third which makes itself known to us by the motion of the equinoctial point, or we might say, by the path of the Sun. This third world indeed remains about as unknown to us as the world of our own willpower is unknown to our ordinary consciousness."

"The Moon however is to be referred to as quite another world, which is, as it were, inserted into ours, and which indicates the breathing process of our Universe, as the Sun indicates the interpenetration of our Universe by the [Cosmic] Ether."

"For one must be quite clear that Sun matter, and any other matter, Earth matter for instance, can under no circumstances be brought into a simple relationship; because *the matter of the Sun is, in comparison with the matter of the Earth, something absorbing and sucking, while the latter exerts pressure. The motions which express themselves in nutation are motions proceeding from the astral world,* and not from anything that can be found in Newton's principles."

Lecture V

"This, however, obliges us to conclude that we cannot speak of a revolution in an elliptical path of the Earth around the Sun, for if the attraction be mutual, we cannot have a one-sided motion of the Earth around the Sun, but both must revolve round a neutral point. In other words, this revolution cannot take place in a manner that would allow us to look on the Sun's center as the pivot, but the pivot must be a neutral point situated between the center of the Sun and the center of the Earth [the barycenter of the solar system]."

"And in this we have to do not only with an aggregation (in the Sun) of this absorbent ether in the outer Universe, but also with the fact that this *ether is distributed far and wide, everywhere we find, coexisting with the force of pressure, the absorbent force. We ourselves carry this force of suction in our own etheric bodies.*"

"With this we completely exhaust all that we call Space. Pressure and Suction—these two, we find in Space. But not only do we possess our physical body, composed of ponderable matter, which it assimilates and again expels, we also have an etheric body, composed of absorbent ether, but we have in addition an astral body—if we may use the term 'body' in this connection. What does the possession of this third body imply? *It means that we have within us something [astral body] that is no longer spatial, though it has a certain relation to space.* This relationship can be proved when we realize that during waking hours the astral body interpenetrates the etheric and physical bodies. But the etheric body acts very differently when we are awake and when we are asleep. A different relation is established between the etheric and physical bodies when we wake, and this is caused by the astral body. *It [the astral body] is active, and works upon the spatial, though it is not itself spatial. It brings order and organization into the correlations of space. This organizing activity of the astral body within us takes place also in the outer Universe.*"

"*There simply does not exist such a thing as mathematics calls 'Space'; but everywhere are lines of force, directions of force, and these are not equal, they vary, they are differentiated. He possesses them through his astral body, which establishes these relations.* They are only possible to him because through his astral body Man is a denizen of an astral world, a world which, though acting upon Space, is not itself spatial. *We only conceive the Zodiac in its full meaning when we treat it as the representative of the astral world beyond.*"

"We know that either actually or apparently, the Sun passes through the zodiac in various ways, in its daily course, in its yearly course, and again in its course through the Platonic Year, through the Precession of the Equinoxes. This points to the fact that the effects upon us of that absorbent ether ball called the Sun vary greatly, as they come from different directions of space."

"*We may therefore consider the astral body as being in a sense part of the celestial, and not of the terrestrial Universe*, for when, together with the Ego, it is outside the physical body, we have to co-ordinate it to the non-terrestrial influences."

"Mankind, like a child, is always more or less asleep, therefore the child is much more receptive to the celestial influences than the adult."

"During childhood, all that is within the skin is still plastic and in a state of formation. The formative powers become less and less active with the years, until, at a considerably later point in life, they become very small indeed. This shows that the inner physical formation-process stands in a certain relation to the movements and configurations to the outer celestial Universe."

"*With all that is in him through his astral body and Ego, man is related to forces that are active beyond the Earth.*"

"*The human form is modeled by forces residing in the periphery and they are forces from beyond the Earth.* Therefore, when we contemplate a human form, we are looking at a product of the extra-earthly."

"Secondly, we can distinguish in Man, apart from his form, all that comes under the category of internal motion. Take, for instance, blood and other bodily juices; these possess internal motion. This also is produced from within; it is, so to speak, situated even deeper in Man than his form. The latter presses forward to the periphery, while internal motion takes place entirely within; and it is again a process that stands in relation with the world that is beyond the Earth."

"Thirdly, the activity of the organs. Organs such as the lungs, liver, spleen, etc., are responsible for activities within Man, and it is these activities I will name as the third thing we find in Man. Although the organs, in the manner of their formation, depend upon the flow of inner forces, yet they are something in the inner part of Man that again comes into relation with what is outside. We have here something which, as you can see from an example I will give, stands in closer relation with the earthly; through these organs, we are brought from the interior into contact with the exterior. You see, *all these processes that take place entirely within us, the formation and regulation of fluidic movements, etc., stand in a relationship with the extra-earthly; whereas when we come to the organs, we again approach the terrestrial. Here we have the union of Heaven and Earth in Man.*"

"Thus, we can study man from four different points of view: Complete Form, in so far as this is built up from within outwards, Internal Motion, Organic Activity, and Metabolism. If we study the complete form, which is entirely constructed by inner forces, we find that it has the least connection of all with the earthly. We only begin to gain an understanding of the connection when we relate, as we shall do tomorrow, the complete form of Man to the Zodiac. The inner motion, the circulation of the blood, lymph, etc., can only be conceived in their reality, when related to our planetary system. And when we come to the activity of the organs, we are already approaching the terrestrial."

"In this way we can discover Man's relation to the Universe that surrounds him:

Zodiac	(1) Complete form
World of the Planets	(2) Internal Motions
World of the Elements	(3) Activity of the Organs
Earth	(4) Metabolism

"And now consider, if we understand the form of Man in all its nature and conditions, and find the possibility of tracing it back to the Zodiac—that is, to the world of fixed stars—then and then only are we able to form, from Man, an idea of *all that is visible to us in surrounding space; for it cannot be investigated by mechanical or mathematical means, but only through a knowledge of the complete form of Man.* Neither are planetary motions to be examined merely by means of a telescope. With a telescope one finds their positions—setting it first to one star and then to the other, finding the angle, and in this way discovering the positions. What is actually present in the processes of the Planet-World is something that is formed from within outwards. *It is by a study of the activities in the saps and juices in Man that we shall learn to understand the planetary activities.* Similarly, if we comprehend our own organic activities, we shall also understand what goes on in the Elemental world; and when we are able to understand what happens in Man in the moment when earthly substance is introduced into his metabolic system, we shall possess the key to the Earth activities and be able to separate them spatially from all extra-earthly activities."

Lecture VI

"To begin with, we must consider all that has connection with the formative forces. These are the forces which work outward from within until they reach the outermost periphery, the limits of man's circumference. The formative forces must have a connection with the activity of the senses."

"The relation of metabolism in Man to the forces responsible for the form of Man is the relation of the Earth to the Heaven of Fixed Stars, which latter is represented for us by the Zodiac."

"When we look at the Zodiac, it forms for us the outer cosmic representative of our own outer form."

"The 'movements' include that of the blood as well as the nerve-fluid [cerebral spinal fluid], lymph, etc. We need not give a detailed list of them here, but *there are seven of these inner movements.* Connected with these movements are the individual organs. The forces of motion have produced, within their courses, these organs; the pulsation of the heart is not the cause but the effect of circulation. *Into the living inner motions or movements is inserted the functioning of the organs.*"

"The heart must be viewed as a creation of the circulation of the blood. We must, so to speak, insert the heart into a living blood-circulation. The movement of the Sun, too, must be thought of as similarly inserted into the movements of the Planets."

"...Earth and Sun move together in the course of the year; the one does not circle around the other."

"It is absolutely necessary to seek in the movements of the outer Universe the equivalent of movements taking place within Man. For only by a study of the changes within physical Man can we arrive at an understanding of the planetary motions that are exterior to Man."

"That which we have termed the organ-building forces, active in the course of the year, are the equivalent of the annual motion of Earth and Sun together."

"The Earth moves behind the Sun in a screw-like line, the Earth moving along always with the Sun. And then, if we view the line from above, we get a projection of the line, and the projection shows a lemniscate."

"Now, all this will make it clear that we can certainly speak of a daily motion of the Earth around her axis, but by no means of a yearly motion of the Earth around the Sun. For the Earth follows the Sun, describing the same path."

"Various other facts show that we have no right to speak of such a revolution. That line which should be apparent to us, and which would correspond to the progressive motion of the Earth in her relation to the Sun, is not to be found [in scientific calculations or theory]."

"It is in a spiral, screw-like path that the Earth follows the Sun, boring her way, as it were, into cosmic space."

"I have already indicated, however, that there is another movement which manifests in the phenomena of the Precession of the Equinoxes—the movement of the point of sunrise at the Spring-equinox through the Zodiac, once in 25,920 years. When the vernal equinox has come back again to the same place, the alterations that will have taken place in the whole human race are such that the human form will be quite dissimilar to what it is now."

"...Man, as he now is, built up in accordance with the forces of the periphery and of inner movement, with the forces too of the organs and of the metabolic system, is complete and finished in his dependence on the forces from without; and now he is able, with his complete and finished organization, to sever himself from this connection."

"Man's present experiences are fundamentally a picture or copy of his past adaptation to the Macrocosm, and in this sense, we live in the pictures of our past. Within these, we are enabled to evolve our freedom, and from them we receive our moral laws, which are independent of the necessity ruling in our nature. It is when we understand clearly how Man and Macrocosm are related to each other that we recognize the possibility of free-will in Man."

"The ancient wisdom therefore spoke of the Zodiac or Animal Circle, not of the Man Circle, as corresponding to the forces of formation. The forces of form manifest themselves in the animal kingdom in a great variety of forms, while in Man they manifest essentially in one form covering the whole human race; but they are the forces of the animal kingdom, and as we evolve beyond them and become Man, we must go out beyond the Zodiac. Beyond the Zodiac lies that upon which we, as human beings, are dependent in a higher sense than we are upon all that exists within the Zodiac, that is, within the circle of the fixed stars. *Beyond the Zodiac is that which corresponds to our Ego.*"

"With the astral body—which the animal also possesses—we are fettered to a dependence upon the Macrocosm, and the building up of the astral vehicle takes place in accordance with the will of the Stars. But with our 'I', or Ego, we transcend this Zodiac."

"Here we have the principle upon which we have gained our freedom. Within the Zodiac, we cannot sin any more than can the animals; we begin to sin as soon as we carry our action beyond the Zodiac. This happens when we do that which makes us free from our connection with the Universal forces of formation, when we enter into relationship with regions exterior to the Zodiac or region of fixed stars. And this is the essential content of the human Ego."

"You see, we may measure the Universe in so far as it appears to us as a visible and temporal thing, we may measure its full extension through space to the outermost fixed stars, and all that takes place by way of movement in time in this starry heaven, and we may consider all this in its relation to Man; but in Man is being fulfilled something that goes on outside this space and outside this time, outside all that takes place in the astral. There beyond, is no 'necessity of Nature', but only that has a place which is intimately connected with our moral nature and moral actions. Within the Zodiac, we are unable to evolve our moral nature; but in so far as we evolve it, we record it into the Macrocosm beyond the Zodiac. All that we do remains and works in the world. The processes taking place within us, from the forces of formation to the forces of metabolism, are the result of the past. But the past does not prejudge the whole of the future, it has no power over that future which eventuates from Man himself in his moral actions."

Lecture VII

"For instance, materialism does not realize that our internal organs have very much more to do with our memory than has the brain, which merely prepares the idea or images so that they can be absorbed by the other organs of the whole body. Here it is that memories originate through the transmutation of the organs."

"The descriptions of the Universe circulated since the beginning of modern times are mere abstractions, for they are described without a knowledge of Man."

"This is the reform that is necessary, above all, in Astronomy —a reform demanding the re-inclusion of Man in the scheme of things when cosmic movements are being studied. Such studies will then naturally be somewhat more difficult."

Lecture VIII

"You know that the intimate relation between man and the Universe was much better known to methods of perception used by the ancients than to ours of the present day. But in those ancient times, by means of course, of atavistic perception, Man knew himself to be dependent also in his soul on the one hand on the elements of fire, water, and air, and on the other hand on the movements of the planets. These he related to his soul nature in the same way as he related the products of the Earth to his physical metabolism. And the part of the Universe that is outside or beyond the planetary system, all that is in the starry heavens, this he connected with his spirit."

"For it is absolutely impossible to gain a true perception even of the visible starry heavens unless man combines with the outer physical life the super-physical in his considerations—that super-physical part of his life through which he passes between death and re-birth."

"Conception is merely the opportunity given for a certain cosmic activity to penetrate the human body, and that which is formed first in the process of man's formation is indeed an image of the whole Cosmos. *He who wishes to study the human embryo from its first stage onwards must consider it as an image of the Cosmos.*"

"The truth is that the minutest molecule is acted upon by the whole starry heavens. The Planets also transmit their

forces in the same way, and we come to realize that the molecule is nothing but a focus of macrocosmic forces."

"Out there in the Macrocosm, we have to behold the structure of the molecule that is only reflected in the molecule. Instead of passing into the infinitely little, we must turn our gaze outwards to the infinitely great, for it is there we must look for the reality of what lives in the little. Neither is the materialism, relative to humanity, overcome so long as we continue to ignore the connection of Man the Microcosm with the Macrocosm."

"The truth is that *in order to arrive at a real understanding of the world, we must conceive of heavy, ponderable matter as ceasing at the ether; for we must clearly understand that this ether is essentially a very different thing from that substance of which we speak as filling space.* When ether approaches an object, it attracts it and draws it within itself. *The activity of ether is the exact opposite to that of matter.* Ether acts as an absorbent."

"We have on the one hand, the pressure of ponderable matter, and on the other, the absorbing action of ether. Your nose is forced outwards, as it were, from your organism through the pressure of matter, while the eye sockets are drawn inward through the action of ether. It is therefore simply a pressing and absorbing substance acting within you which differentiates the exterior appearance of your front and back."

"From the foregoing, you will now see that it is necessary, when speaking of the life between death and rebirth, to realize that this life is actually lived beyond space—beyond the space of which we are cognizant on the Earth-plane; and we shall have to endeavor to gain a conception of this 'beyond' of space."

"*That which mediates also in you between pressure and suction, is the astral.* No relation would exist between the front and back of a human body did not the astral activity within form the connection between the absorbent and the pressing elements."

"The astral is active there, and its activity is based upon the contrast between the front and back nature of the human being, even as the connection between the higher (head) man and lower (limb) man by way of the astral is based upon the Ego."

"Empty space can be less than empty by being filled with ether, when it becomes a negative entity. And that which adjusts or connects the two, that which mediates also in you between pressure and suction, is the astral."

"We must therefore consider man, as he stands before us, in a quite concrete manner and make clear to ourselves that while he has existence upon this plane *between birth and death, he imprints his astral part and his Ego in the absorbent and pressure-producing elements*, but his being only manifests here on Earth as the mediator between the front and the back, and between the upper and the lower parts of the body."

"The only way we experience the activity of the Earth's revolution upon its axis is in our daily metabolic process. We must take our daily meals, and this, together with the succeeding digestive processes, takes place within the limits of 24 hours, uniform with one revolution of the Earth. These two things belong together, the one is proof of the other. When we die, the revolution of the Earth becomes something real, as real as are the visible objects here. Then we live with this terrestrial motion; we begin to experience this motion consciously."

"Between birth and death, we say: My heart is within my breast, and in it converge the streams or motions of the blood-circulation. At a certain stage of development between death and re-birth, we say: In my inner being is the Sun—and by this expression we mean the actual Sun, which the physicist claims to be a ball of gas, but which is in reality something quite different. We experience the actual Sun in the same manner as we experience here the heart. Here the Sun is visible to the eye, whereas during the time between death and re-birth it is not. The evolution of the heart on its path to the pineal gland, as it undergoes on the way a wonderful metamorphosis, is the cause of sublime experiences. The complete system of our blood-circulation we experience consciously in its transformation; we have this system within us as existence between death and rebirth proceeds, these forces undergo transmutation, so that, when once again we arrive at the gates of a new Earth-life, they have become the forces of us—not, of course, the substance, but the forces."

Lecture XI

"Thus, we may say that in Venus and Mercury we have something which places us outwardly, as a member of the planetary system, in relation to a super-sensible world. Here, while we live, we are brought into connection with a super-sensible world through Venus and Mercury. We might say: When we are incorporated by birth into the physical world, we are carried into it by Saturn and Jupiter; while we live from birth to death, Venus and Mercury work within us and prepare us to carry our super-sensible part back again through death into the super-sensible world. In fact, Mercury and Venus have just as much share in our immortality after death as Jupiter and Saturn have in our life before death. It is really so that we have to see something in the Cosmos which corresponds to the relation between the comparatively more spiritual organization of the head and the rest of the human organization."

"For if we take Saturn, the outermost planet of our system, we must represent him as the leader of our planetary system in cosmic space. He directs our system in space. He is the body for the outermost force which leads us round in a lemniscate in cosmic space. He is the driver and the horse at the same time. Saturn is thus, the force in the outermost periphery. But there are other forces in our planetary system which show a more intimate adjustment to the spiritual world—the forces that we find in Mercury and Venus. Through these forces, our path is continually raised. Thus, when we look upon the path from above, we have the lemniscate, but when we look at it from the other side, we obtain lines which are continually rising upwards; there is a progression."

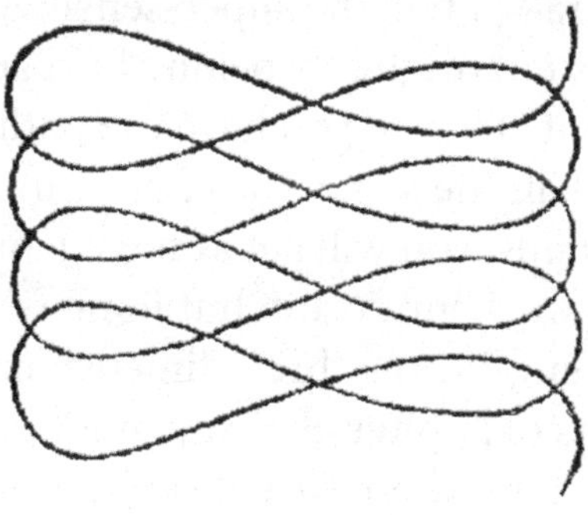

"This progression corresponds in man to the fact that during sleep what we have taken into us, though it may not pass over at once into consciousness, is elaborated; during sleep, we work upon it. It is principally during sleep that we work on what we have absorbed through our life, our training, and education. During sleep, Mercury and Venus communicate that to us. They are our most important night planets, as Jupiter and Saturn are our most important day-planets. Hence the old instinctive atavistic wisdom was right in connecting Jupiter and Saturn with the formation of the human head, and Mercury and Venus with the formation of the human trunk, and the rest of the organism. These things arose from an intimate knowledge of the connection between Man and the Universe."

"Now I will ask you carefully to consider the following. It is first of all necessary to understand, from inner grounds, the movement of the Earth. We must recognize the influence upon it of the Venus and Mercury forces, which themselves bear the lemniscate on further, so that it progresses, and its axis becomes itself a lemniscate. We have thus, for the Earth, an extremely complicated movement."

"Planets such as Venus and Mercury, however, have a relation to the extra-spatial, the super-sensible, the spiritual, to that which does not originally belong in space, but has, as it were, come into it. Thus, if you have the paths of Saturn, Jupiter, Mars, and, in the same space, draw in also the paths of Mercury and Venus, you will get at most a projection of the Mercury or of the Venus orbit, but in no sense the orbits themselves. If we employ the three-dimensional space to sketch in the orbits of Jupiter, Saturn, and Mars, we come, at most, to a boundary where we get something like a path of the Sun. But if we wish to draw the others, we can no longer do so in the three-dimensional space, we can only get shadow-pictures of these other movements in it; we cannot draw the path of Venus and that of Saturn in the same space. From this, we see that all delineations of the solar system where the same space is used for Saturn as for Venus are only approximate, they do not suffice for a solar system."

"Just as in the case of man, in order to understand him as a whole, we must pass from physical to super-sensible forces; so, in the same way, *to understand the solar system, we must pass from the three dimensions into other dimensions. That is to say, we cannot delineate the ordinary solar system in the three-dimensional space.*"

Lecture XII

"Ultimately, astronomy cannot really be studied at all without entering into a study of the whole being of Man."

"The Sun and Moon are not from the same origin but are two streams running side by side; and just as little can Man's human ego and astral body be traced to the same origin as his physical and etheric bodies. They are two different streams."

"We have seen that if he places himself rightly, the result is not one astronomy, but two; a solar and a lunar astronomy. So too, we have a human development of a heathen nature—natural science is still heathen—and a human development of a Christian nature."

Lecture XIII

"It is not possible to understand the Universe in itself without keeping in mind Man and the relation of the Universe to him."

"The watery element in Man has, however, primarily no connection with the Earth but with the planetary Universe outside, and especially with the Moon."

"I described yesterday that we have on the one side the astronomy that applies to the Sun—and also to the Earth. We ourselves are part of that astronomy, for we are organized into it as organisms containing solid substances. Lunar astronomy, however, is different. We are organized into lunar astronomy in so far as it is connected with our fluid constituents."

"The whole connection in the Universe is so constructed that it sustains a man's whole life for a solar day, which is 72 years—a solar day."

"Man is so connected with the Cosmos that the duration of his life is one day of the cosmic year. He was thus organized into the Cosmos."

"For science must not simply continue to develop, as did the science of the last century, under the influence of the materialistic view of the Universe, which is itself a child of the materialistic comprehension of Christianity. We owe materialism to the materialization of the Christian view of the Universe. The teaching of the cosmic Christ must be re-established in opposition to the materialized form of Christianity we have today. This is the most important task of our time; and until its importance is realized, man will not be able to see clearly in any domain."

Lecture XIV

"We have now come to a time when what we may call the Cosmogony of Natural Science, and what we call the Moral Order of the Universe—to which, of course, also belongs the religious view of the world—stand side by side, utterly unconnected."

"*Man is born out of time; he is so born that he owes his existence to the difference in velocity between the cosmic day of the stars and the cosmic day of the Sun.* In modern parlance, we should say: If the Sun were not in the Universe as it is, if it were just a star like other stars, having the same velocity as other stars, what would be the consequence? It would be that the luciferic powers alone would rule. That this is not so, that man is able to withhold himself from the luciferic powers with the whole of his being, is due to the circumstance that *the Sun does not share in the velocity of the stars but lags behind them,* not developing the luciferic velocity but the velocity of Jehovah. Again, if there were only the Sun velocity and not that of the stars, man would not be able to run on in front of the rest of his development with his mental powers, as he does at present. Understanding travels with the stars, experience with the Sun."

"The solar movement is outside the system to which the Moon and stars belong. Thus, *we are so situated in the Universe that on the one hand we are coordinated to the stellar-lunar system, and on the other to the solar movement.*"

"On the one hand, we have the starry system which, in a certain respect, contains within it the Moon; and on the other, the system to which the Sun belongs. They mutually interpenetrate. They work together. But we are wrong if we apply the same law to the two."

"The cosmic happenings in which we are involved have two origins, but we are so placed that these two streams flow together in us. They fuse in us human beings."

"What we have as a fact of consciousness is the content of our soul-life. That can exist under no other hypothesis than that the whole of this quite material happening is demolished, is annihilated, and that in us there actually is no conservation of force and substance but room is made by the annihilation of substance for the development of the thought life. In fact, Man is the one arena in which an actual annihilation of substance takes place."

"There is in us a continual tendency to work with a lesser velocity over against the more rapid velocity of the stars. The action of the stars in us is opposed."

"Here, let us say, a certain movement occurs (arrow pointing downwards), i.e. a flow takes place with a certain velocity; and the stream then fuses with another stream—it must be assumed that the other flow is going not in the same but in the opposite direction (arrow upwards)."

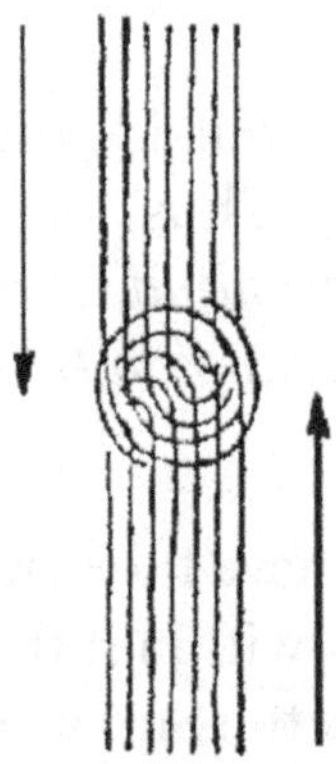

"The two streams flow, therefore, into one another. Or imagine a wind whirling at a certain velocity from above downwards, and another from below upwards, and they whirl into one another. If we take the difference of velocity between the downward and the upward current, relating the latter to the former in such a way that a difference in velocity results bearing the same relationship as the difference in velocity between the stellar time and the solar time, then through the rotation a condensation arises which receives its own distinct form. *One whirls downwards, and because the other whirls upwards driving with a greater velocity, the lesser velocity would be that driving downwards, which gives here (see diagram) through the collision, a condensation, a certain*

figure. This figure, disregarding imperfections, is a silhouette of the human heart."

"Thus, through the meeting of the Lucifer stream and the Jehovah stream, it is possible to construct exactly the figure of the human heart. It is constructed simply out of the revelations of the Universe. It is absolutely true that *the Sun-movement is an expression of a slower movement which meets a quicker movement, and we are so inserted into the two movements that the silhouette of our heart arises; and on to it the rest of the human form is fitted.* We see from this what Mysteries are actually hidden in the Cosmos, for as soon as we admit *we have two astronomies*, which work together in their results—what is the result? The human heart."

Lecture XV

"The human brain stands in relation to the starry sphere, and human eyes are in a certain way coordinated with the Sun-sphere."

"We have now apportioned man, taking him in his totality, to the starry world, the solar and planetary world, and the Earth-world. We have placed him into the whole Universe and when we look at his brain inwardly, not outwardly; not by physical anatomy, but by inner knowledge—we *see in the human head*, inasmuch as it is the bearer of the brain, *a direct copy of the starry world. We see in all that extends from the nose to the lungs, a copy of the planetary system with the Sun.* If we then consider *the remainder, we see that part of man which is Earth-bound, as e.g. are animals.* In this way only do we arrive at the true parallel between man and the rest of the world."

"Just as the circulatory course of the lungs with its absorption of oxygen and giving out of carbonic acid belongs to the *planetary system*, just as what is introduced through the digestive apparatus belongs to the *Earth*, so that part of the circulatory course that branches off above, belongs to the *starry world*."

"The human brain is nothing apart from the starry heavens."

"When looking at the Sun, scientists imagined they would find there a glowing ball of gas. They would not find this, but a suction-sphere, trying to absorb everything possible into itself, really an empty space, nay, even less than empty, a negation of matter. *Within the circumference of the Sun, there is nothing comparable to our "matter". It is not merely empty, but less than empty; it is blank, just like a hole, in comparison with the rest of matter*."

Lecture XVI

"When we see a man before us, we are also confronted by an enclosed heat-space, which is at a higher temperature than its environment. *In this increased temperature lives the soul and spirit element of man*, and the soul and spirit in him is indirectly conveyed by means of the heat, to his other organs. In this way too, man's willpower comes into existence."

"This process, as a matter of fact, takes place in everything that may be called the transition to pure thinking. All material substance which has cooperated with human inner life falls back into the organism and pictures alone remain. It is a fact that *when we rise to pure thought, we live in pictures*. Our soul lives in pictures; *and these pictures are the remains of all that has gone before*. Not the substance, but the pictures remain."

"What has just been presented can be followed into the thoughts themselves, for this process only takes place at the moment when thoughts change into mere pictures. At first, thoughts live, as it were, embodied. They are permeated by substance; but as pictures, they separate themselves out from this substance. If, however, we go to work in a truly spiritually scientific way, we can quite easily distinguish pure thought, sense-free thought that has separated itself out from the material process, from all thoughts belonging to what I have called in these lectures the "instinctive wisdom of the ancients."

"The thoughts of earthly humanity before the Mystery of Golgotha were always united with matter, and that at the time of the intervention of the Mystery of Golgotha in Earth-life, humanity had arrived at the point in evolution of being able to dissociate itself from matter in the inner process of thought; matter-free thought became possible."

"Man in his evolution has become free from the embodiment of thoughts; that they have changed to pure pictures. Thus, we may say that up to the time of the Mystery of Golgotha, embodied pictures lived in man; but after the Mystery of Golgotha, matter-free pictures lived in man. Before the Mystery of Golgotha, the Universe worked upon man in such a way that he could not attain to pictures free of the body, free of matter. Since the Mystery of Golgotha, the Universe has, as it were, withdrawn. Man has been transposed to an existence which only takes place in pictures."

"In man alone, matter is thrown out by pure thought. That matter which is actually cast out of the human being by pure thought is also annihilated, it passes into nothingness."

"Here is the Earth, and on the Earth, man; into man passes matter. Everywhere else, it is transmuted. In man, it is annihilated. The material Earth will pass away in proportion as matter is destroyed by man."

"When, someday, all the substance of the Earth will have passed through the human organism, being used there for thinking, the Earth will cease to be a cosmic body. And what man will have gained from this cosmic Earth will be pictures. These, however, will have a new reality, they will have preserved an original reality. This reality is that which proceeds from the force which, as central force, makes itself felt through the Mystery of Golgotha."

"The end of the Earth will come when all its substance is destroyed as described above. Man will then possess pictures of all that has taken place in earthly evolution. At the end of the earthly period, the Earth will have sunk into the Universe, and there will remain merely pictures, without reality. What gives them reality, however, is the fact of the Mystery of Golgotha having been there in humanity; that gives these pictures inner reality for the life to come. Through the Mystery of Golgotha, a new beginning is set for the future existence of the Earth."

"From this, we can see that what is contained in our stream of evolution is not to be regarded merely as a continuous stream, where one thing is always related to another as effect to cause, but we must so consider the Earth-evolution that we recognize, in the first place, a pre-Christian evolution, out of which came all that men were able to think at that time, for what they were able then to think was contained in the Father-God, was imparted to the Earth through Him. The nature and work of the Father-God, however, was such that what He created as Earth-evolution was given over to that part of Earth-evolution that tends to pass away. A new beginning was made with the Mystery of Golgotha. Of all that went before, only pictures were to remain, as it were, descriptive paintings of the world. These pictures were, however, to receive new reality through that which entered as Being into the evolution of the Earth through the Mystery of Golgotha. That is the cosmic significance of the Mystery of Golgotha; that is what I meant years ago, when I said: *Christianity will not be understood until it has penetrated even into the physics of our Earth, until we understand how, even in physical things, the Christian substance works in the world-existence.* We have not grasped Christianity until we can say to ourselves: Precisely in the domain of heat, such a change is taking place in man that through it matter is being destroyed and a purely picture-existence comes forth out of the matter; but through the union of the human soul with the Christ-substance, this picture-existence is made into a new reality."

"Man, however, retains these pictures, and they derive a reality for the future because a new substance has laid hold of them, the substance which has passed through the Mystery of Golgotha."

"All proofs of freedom fail; our task is not to prove freedom, but to take hold of it. It is grasped when one understands the nature of sense-free thinking. Sense-free thinking, however, needs again the connection with the world, and this connection it does not find unless it unites with what has been introduced into the evolution of the world as new substance through the Mystery of Golgotha."

"It might at first appear very strange that just those who uphold the modern creeds—as well as ancient creeds that extend their influence into modern life—do not desire a science leading towards Christianity, but desire a science as materialistic as possible, so that *an unscientific faith* may hold its own alongside of it."

"When we come as far as Heat, we find the transition to the soul and spirit nature, for *in Heat, we have already the transition from Space to Time*; and that which is of the soul flows in the temporal. *Beyond Heat, we pass more and more out of Space into Time, and it becomes possible, by the roundabout way here indicated, to seek the moral in the physical.*"

"Today, the method of study is such that we look up into abstraction, we have our thoughts up there; and we look down into the physical material. We do not make the transition unless we pass over to the inwardly stirring heat lying between these, which has, at least for human instinct, still a physical as well as a soul aspect. We can develop warmth for our fellows morally—soul-warmth, which is the counterpart of physical warmth."

"What today is only soul-warmth will in a future cosmic existence play a physical part, for the Christ-impulse will live therein. What today is simply picture-warmth in our world of Feeling—will live on, that it may become physical when the Earth-warmth has disappeared, for it is what the Christ-substance, the Christ-nature, is. Let us try to find that delicate connection between external physical warmth and that which we instinctively call warmth of feeling..."

The Interior of the Earth

The following descriptions of the interior of the Earth are taken from *An Esoteric Cosmology*, Lecture XVI, *Earthquakes, Volcanoes, and Human Will*, Rudolf Steiner, Paris, June 12, 1906, GA 94.

"Physical science as yet only knows of the terrestrial crust, a mineral layer which in fact is only like a thin skin at the surface of the Earth. In reality, the Earth consists of a succession of concentric layers which we shall now describe:

1) The mineral layer contains all the metals which are found in the physical bodies of everything that lives at the surface. This crust is formed like skin around the living being of the Earth. It is only a few miles in depth.

2) The second layer can only be understood if we envisage a substance which is the very opposite of what we know. It is negative life, the opposite of life. All life is extinguished there. Were a plant or an animal plunged into it, it would be destroyed immediately. It would be totally dissolved. This second shell—half liquid—which envelopes the Earth is truly a sphere of death.

3) The third layer is a circle of inverted consciousness. All sorrow appears there as joy. And all joy is experienced as sorrow. Its substance, composed of vapors, is related to our feelings in the same negative manner as the second layer is regarding life. If we now abstract these three layers by means of our thinking, we will then find the Earth in the condition in which it was before the separation of the Moon. If one is able by means of concentration to attain a conscious astral vision, one would then see the activities in these two layers: the destruction of life in the second and the transformation of feelings in the third.

4) The fourth layer is known as water-earth, soul-earth, or form-earth; it is endowed with a remarkable property. Let us imagine a cube and now picture it reversed inasmuch as its substance is concerned. Where there was substance, there is now nothing:

> the space occupied by the cube would now be empty while its substance, its substantial form, would now be spread around it, hence the term "earth of form." Here this whirlwind of forms, instead of being a negative emptiness, becomes a positive substance.
>
> 5) This layer is known as the earth of growth. It contains the archetypal source of all terrestrial life. Its substance consists of burgeoning, teeming energies.
>
> 6) This fire-earth is composed of pure will, of elemental vital forces—of constant movement—shot through by impulses and passions, truly a reservoir of will forces. If one were to exert pressure on this substance, it would resist.

If now again in thought one were to abstract these last three layers just described, one would arrive at the condition in which our globe was when Sun, Moon, and Earth were still interwoven.

The following layers are only accessible to a conscious observation which is not only that of dreamless sleep but a conscious condition in deep sleep.

> 7) This layer is the mirror of the Earth. It is similar to a prism which decomposes everything that is reflected in it and brings to expression its complementary aspect, seen through an emerald, it would appear red.
>
> 8) In this layer everything appears fragmented and reproduced to infinity. If one takes a plant or a crystal and one concentrates on this layer, the plant, or the crystal, would appear multiplied indefinitely.

9) This last layer is composed of a substance endowed with moral action. But this morality is the opposite of the one that is to be elaborated on the Earth. Its essence, its inherent force, is one of separation, of discord, and of hate. It is here in the hell of Dante that we find Cain, the fratricide. This substance is the opposite of everything which among human beings is good and worthy. The activity of humanity in order to establish brotherhood on the Earth diminishes the power of this sphere. It is the power of Love which will transform it inasmuch as it will spiritualize the very body of the Earth. This ninth layer represents the substantial origin of what appears on Earth as black magic, that is, a magic founded on egoism.

"These various layers are connected by means of rays which unite the center of the Earth with its surface. Underneath the solid earth there are a large number of subterranean spaces which communicate to the sixth layer, that of fire. This element of the fire-earth is intimately connected with the human will. It is this element which has produced the tremendous eruptions that brought the Lemurian epoch to an end. At that time, the forces which nourished the human will went through a trial which unleashed the fire catastrophe that destroyed the Lemurian continent. In the course of evolution, this sixth layer receded more and more toward the center and, as a result, volcanic eruptions became less frequent. And yet they are still produced as a result of the human will which, when it is evil and chaotic, magnetically acts on this layer and disrupts it. Nevertheless, when the human will is devoid of egoism, it is able to appease this fire. Materialistic periods are mostly accompanied and followed by natural cataclysms, earthquakes, etc. Growing powers of evolution are the only alchemy capable of transforming, little by little, the organism and the soul of the Earth.

"The following is an example of the relationship that exists between the human will and telluric cataclysms: in human beings who perish as a result of earthquakes or volcanic eruptions one notices, during their next incarnation, inner qualities which are quite different. They bring from birth great spiritual predispositions because, through their death, they were brought in touch with forces which showed them the true nature of reality and the illusion of material life.

"One has also noticed a relationship between certain births and seismic and volcanic catastrophes.

"During such catastrophes, materialistic souls incarnate, drawn sympathetically by volcanic phenomena—by the convulsions of the evil soul of the Earth. And these births can, in their turn, bring about new cataclysms because, reciprocally, the evil souls exert an exciting influence on the terrestrial fire. The evolution of our planet is intimately connected with the evolution of the forces of humanity and civilizations."

1. Mineral crust
2. Negative life
3. Inverted Consciousness
4. Circle of forms
5. Circle of growth
6. Circle of fire
7. Circle of decomposition
8. Circle of fragmentation
9. Egocentric egoism

The Spiritual Beings in the Heavenly Bodies

***Spiritual Beings in the Heavenly Bodies and in the Kingdoms of Nature*, Rudolf Steiner, Lecture V, April 7, 1912, Helsinki, GA 136**

"When we look at a planet, its form expresses to us the work of the Spirits of Form. Just as behind the human form, behind the physical body, are the etheric body, astral body, sentient soul, etc., so, behind the planet, belonging to it, are what we know as the Spirits of Motion, Wisdom, Willpower, Cherubim, and Seraphim, etc. If we wish to visualize the complete being of a planet in the sense of Spiritual Science, we must say:—The planet eats our perception in cosmic space when its physical being, given by the Spirit of Form, shines forth; and it conceals from us, just as man conceals his higher principles from the physical gaze, all that rules within and around it, as beings of the higher hierarchies. Thus, we rightly imagine such a planet as Mars, or Mercury, only when we first of all picture it in its physical form, and then think of it as surrounded and permeated by a spiritual atmosphere stretching out into infinity and having in the physical planet its physical form, the creation of the Spirit of Form—and as its spiritual environment the beings of the hierarchies. Only when we consider it thus do we conceive of *the complete planet as having the physical as a kernel in the center, and round it, the spiritual sheaths which consist of the beings of the hierarchies.*"

"The external form of the planet is the creation by the Spirits of Form. The inner livingness is regulated by the beings we call the Spirits of Motion or Movement. Now, to the occultist, such a planet is in every way an actual being, a being regulating what goes on within it, according to thought. Not only is that which has just been described as inner vitality present in the planet, but the planet, as a whole, has consciousness, for it is indeed a being. This consciousness, which corresponds to the consciousness of man, to the lower form of consciousness—the subconsciousness in the astral body—is regulated in the planet by the Spirits of Wisdom. We may therefore say: The lowest consciousness of the planet is regulated by the Spirits of Wisdom. In thus describing the planet, we still refer to the planet itself. We look up to the planet saying: *It has a definite form, that corresponds to the Spirits of Form; it has an inner mobility, that corresponds to the Spirits of Motion. All this is permeated by consciousness, which corresponds to the Spirits of Wisdom.*"

"*That which leads the planet through space, which governs its movement through space and causes it to revolve around the fixed star, corresponds to the Spirits of Willpower, or Thrones.*"

"*The harmony of movement between one planet and another corresponds to the activity of the Cherubim.* The regulation of the combined movements of the system is the work of the Cherubim. Each planetary system with its fixed star, which is in a sense the Commander-in-Chief under the guidance of the Cherubim, is again related to the other planetary systems to which other fixed stars belong. These systems mutually arrange their positions in space with due regard to the neighboring systems."

"Wisdom, not acquired in decades, as is the wisdom of eminent men, but such wisdom as is gathered in thousands, nay, in millions of years of cosmic growth, this streams towards us in sublime power from the beings we call the Cherubim."

"In the First Hierarchy, which consists of the Thrones, Cherubim, and Seraphim, we have a form of creation in which the part created is detached—we have there not only self-creation, but world-creation."

"For the inner condition of consciousness, the inner experience of the beings of the First Hierarchy, lies in creation, in forming independent beings. We may say: They contemplate that which they create, and which becomes a world, and it is not when they look into themselves but when they look out of themselves upon the world which is their own creation, that they possess themselves. To create other beings is their inner life; to live in other beings is the inner experience of these beings of *the First Hierarchy. Creation of worlds is their external life—creation of beings their inner life.*"

"In the course of these lectures, we have drawn attention to the fact that these various beings of the hierarchies have offspring; beings split off from themselves, which they send down into the kingdoms of nature, and we have learnt that *the offspring of the Third Hierarchy are the nature-spirits, while the offspring of the Second Hierarchy are the Group-souls.* The beings of the First Hierarchy likewise have offspring split off from them, and as a matter of fact, I have already described from a different aspect these beings which are the offspring of the First Hierarchy. I described them at the beginning of this course, when we ascended to the so-called Spirits of the Rotation of Time, the spirits governing and directing what goes on in the kingdoms of nature in rhythmic succession and repetition. The beings of the First Hierarchy detach from themselves the beings governing the alternation of summer and winter, so that the plants spring up and fade away again; that rhythmical succession through which, for instance, the animals belonging to a certain species have a definite period of life in which they develop from birth to death."

"Everything too which takes place in the kingdom of nature rhythmically and in recapitulation, such as *day and night, alternations of the year, the four seasons of the year, everything which thus depends upon repeated happenings, is regulated by the Spirits of the Rotation of Time, the offspring of the beings of the First Hierarchy.* These Spirits of the Rotation of Time can be described from one aspect, as we did some days ago, and we can now describe them according to their origin, as we have done today. Thus, we can comprehensively represent the beings of these three ranks of Hierarchies as follows:

1st Hierarchy: World-creation	Creation of Beings	Spirits of the Rotation of Time
2nd Hierarchy: Self-creation	Stimulation of Life	Group-Souls
3rd Hierarchy: Manifestation	Being filled	Nature-Spirits with Spirit

"So, that which appears to us when we direct our gaze into cosmic space and there perceive a planet—Mercury, Venus, Mars, or Jupiter—is the external form of the Spirit of Form."

Mutual understanding prevails between one fixed star and another. By this means alone does the Cosmos come into existence. *That which, so to speak, the planetary systems discuss with one another in cosmic space in order to become a Cosmos is regulated by those beings we call the Seraphim.* We have now, as it were, exhausted what we find in man, as far as the Consciousness Soul."

"Just as in man we ascend to his higher spirit-nature, to that which alone gives meaning to the whole system up to the Consciousness Soul, so, if we ascend above the Seraphim, we come to what we tried to describe today as the Highest Trinity of Cosmic Being; we come to that which governs the Universe as the All-pervading, Divine, Threefold Divine Life, which creates for itself sheaths in the different Planetary Systems. Just as that which lives in man as Spirit-Self, Life-Spirit, and Spirit-Man, (Manas, Budhi, Atma), creates sheaths in the Consciousness Soul, Intellectual Soul, Sentient Soul, and the astral, etheric, and physical bodies: so, do the Fixed Stars of planetary systems move through space as the bodies of Divine Beings. Inasmuch *as we contemplate the life of the world of stars, we contemplate the bodies of the gods, and the Divine in general.*"

"We gained an idea of all that participates in forming a planet, and we have seen how a planet receives its form, its enclosed form, as a result of the activity of the Spirits of Form. We saw further that the inner life, the inner mobility of the planet, is the result of the activity of the Spirits of Motion. What we may call the lowest consciousness of the planet, which can be compared with the consciousness present in man in his astral body, we have to allot to the Spirits of Wisdom. That impulse through which the planet, instead of remaining stationary, changes its place in space, we have to allot to the

Spirits of Willpower, or Thrones. The organizing of the planet in such a way that it does not follow an isolated course in space but so moves that its impulses of motion are in harmony with the whole planetary system to which it belongs, the regulating of the individual movements of the planet in harmony with the whole system, that is an activity of the Cherubim. Finally, we ascribe to the Seraphim what we may call the inner soul-life of the planet, whereby the planet comes, as it were, into connection with the other heavenly bodies, as a man by means of his speech enters into relation with other men, so that we must observe a sort of coherence in the planet; and in this, what comes from the Spirits of Form is but a sort of kernel. On the other hand, every planet has something like a spiritual atmosphere, we might even say something like an aura, in which work the spirits belonging to those two higher hierarchies which are higher than the Spirits of Form."

"If you consider the planets—Mars, Saturn, or Jupiter, for instance—which are outside in cosmic space; as you see them with your physical eyes, or with the telescope, you have in the form revealed to you, not merely the Spirits of Form. Let us take, for example, the planet which, for a long period of time, has been reckoned as the outermost one in our system; Uranus and Neptune were added later, as we shall see; but to begin with, we will consider Saturn as the outermost. If we look at Saturn with physical vision, we find him outside in cosmic space, a sort of luminous globe (leaving his rings out of the question). To the occultist who follows the spiritual events in the cosmos, this globe which is seen out there is not what the occultist calls Saturn; to him, Saturn is that which fills the whole space bounded by the apparent elliptical orbit of Saturn."

"Thus, if we seek for the actual place whence the Spirits of Form work, the Spirits of Saturn, Jupiter, Mars, etc., as also the Spirit of Form belonging to our Earth—if we seek for the center, the starting point from which the Spirits of Form work—we find it in the Sun. *That means that the Spirits of Form corresponding to our planets, comprise, as it were, a synod or council of Spirits, having its seat in the Sun*, and from there sets boundaries to certain etheric substances, certain masses of ether, so that what we call occult Saturn, occult Jupiter, comes into being."

"Thus, we have to imagine that where our physical eyes see the planet Saturn, there are two forces working together; the one, the normal force of the Spirit of Form working outward from the Sun; and at a definite point in opposition works the detached Spirit of Form. This produces an "in-turned" structure; the ether is notched, and this notch appears to the physical eye as the physical planet Saturn. Just the same occurs with the physical Jupiter, and with the physical Mars. Hence, by this example, you see how in individual cases there actually arises what we call "maya," the great illusion. Where physical astronomy places a planet, there is in truth a combined working of two forces; and only because, in truth, a great and mighty etheric heavenly body is there, which, through the contact of these opposing forces, is dented in and has a notch formed in one place, does the appearance of the physical planet arise."

"The Spirits of Form working from the Sun extended the etheric substance to a certain distance; there worked the abnormal Spirits of Form in opposition, and caved the substance in, so that in reality, a hollow was made in the etheric substance. As regards the original etheric substance of the planet, where the physical eye believes it sees the planet—there is really nothing; the actual planet is where the physical eye sees nothing. That is the peculiarity of "maya." Where the physical planet is seen, there is a hollow. It may perhaps be said: "It is a very strange idea that where the physical planet is to be seen, there is a hollow," for you will ask about our Earth. In the sense of what has been expounded, our Earth must also be a sort of flattened ball having its central point in the Sun, and it must also be such a notch, such a sort of hollow on the outermost rim. "A fine thing that!" you can say, "We know quite well that we are walking on firm, solid earth. In like manner, we might take for granted that where Saturn, Jupiter, or Mars are, there would naturally have to be solid filling, not hollow. And nevertheless, where you walk about on our Earth—where, in the sense of Maya-perception, you believe yourselves to be walking on solid, firm ground—even then, in reality, you are walking about on a hollow. Our Earth itself, in so far as it is an accumulation of matter, is a hollow in cosmic space, something bored into cosmic space. All physical matter comes into being through the meeting together of forces coming from the Spirits of Form. In this case, we have the meeting of the forces of the normal Spirits of Form and those of the abnormal Spirits of Form. They collide with one another and, in reality, an indentation is produced and consequently at this point a simultaneous breaking up of the form, but only of the form. The form breaks up and this hollow space is bored. Now, broken spiritual form, crushed form, is in reality matter. *In a physical sense, matter only exists when spiritual forms are broken up. Thus, the planets out there are also broken-up forms.*"

"Inasmuch as the beings of the higher hierarchies, the Seraphim and Cherubim, become the bearers of light, they have now the same relation to the light as the forces of the Spirits of Form to the etheric substance. Just as the forces of the normal Spirits of Form pass outwards and encounter the forces of the abnormal spirits working in opposition, and by that means a notch is hollowed out, so also do the forces work which carry the light, filling the whole etheric space; but in opposition to them work the abnormal forces, so that the planet arrests the light. Just as it arrests the forces of the Spirits of Form, so does it arrest the light, and throw it back; hence, it appears as a reflector, as a thrower-back of the light which the spirits we call the Cherubim and Seraphim carry to it from the Sun."

"Hence, the planets have no light of their own because they claim for themselves the force of the light which would be their due as beings if they were to open themselves to the normal Cherubim and Seraphim—because they veil themselves, cut themselves from the whole. Thus, every planet has a cut-off, separated light. *It is not correct to say that the planets only have light borrowed from the Sun; every planet has its own light; but it has cut it off, keeps it hidden within itself, and develops it for its own independent inner life of light. We shall see that each planet only shares this light with its own beings, belonging to the kingdoms of nature* on the planet in question. But that light, to which they ought to open themselves, which they ought to take up from outside, is brought to them from the Sun by the Cherubim and Seraphim, but to that they close themselves, and throw it back. Hence, seen in cosmic space, they are stars which have no light of their own. Thus, as it were, with the light which flows in from the Sun, a notch is formed, and the planet throws itself against that light flowing in from the Sun; arrests it and throws it back."

"Thus, to occult vision, what we observe in the stellar world, is absolutely different from what it appears to physical astronomy. What exists for the latter is nothing but a description of a Maya, and only behind this Maya does the truth lie; for the truth behind the material world is the spiritual world. *In reality, the material world does not exist at all. What is called the material world is the interplay of the forces of the spiritual world.*"

"Thus, what the eye sees is really nothing but the reflected light, which is thrown back, because, *when the Seraphim and Cherubim carry the light of the Sun into cosmic space, the luciferic Cherubim and Seraphim throw themselves against them, so to speak, and insert darkness into the substance of the sunlight*; cutting off the light within, and claiming for each of the planets a light of its own."

Lecture VII

"Occult perception then realizes that when we look up to the fixed star, what we perceive as working in the fixed star is the etheric body of the planetary system; for we actually gain the impression of an etheric body. We understand the impression which the fixed star makes upon us when we observe the etheric body of the plants, where it works as yet unmixed with an astral body, where only the etheric body works, in cooperation with the physical body. We then gain the knowledge that when we look at the fixed star, we actually see the etheric body of the planetary system raying down from it. We can now say: In the moons, we have the corpse of the planetary system, in the totality of the planets, we have its body, its physical body; and in the fixed star itself, raying out from it, we have the etheric body of the planetary system. In fact, the possibility of occult vision clinging to the dead paper-mache ideas upon which all physical astronomy is based, soon ceases; for this vision everywhere recognizes that the whole planetary system is permeated by life and is a living organism. Indeed, a continual stream of etheric life flows from the fixed star to the outermost boundary of the system, and back again.

We have continually to do with life-forces (as in living animal and plant bodies) which appear centered—I say that now by way of comparison—centered in the fixed star—as the life of the animal is centered—let us say—in the heart; or as plant-life is centered in the various organs which regulate the rising and falling of the sap. In short, we have to do with a center of the planetary system, which we must seek in the fixed star."

"If we should only investigate all the harmful forces brought forth by the various luciferic spirits, we should find an enormous mass of harmful, astral substances in a planetary system. And, in a curious way, the occult vision, which for a time has had the opportunity of observing cometary life, shows us that *everything of a cometary or meteoric nature is ever striving to collect round itself all the harmful astral products, and to remove them out of the planetary system.* We shall see further in the course of these lectures how this affects in particular the harmful astral products of humanity; but we see that the noxious luciferic evils are carried away through the comets out of the planetary system."

"Thus, we must regard cometary life as something which continually works as a thunderstorm in the planetary system, cleansing it. By the passage of a comet through the planetary system, an effort is made to discharge from it the injurious matters which have been called forth in the system by the harmful astral radiations of the beings. Thus, we have something in the cometary life of which we cannot, as in the physical and etheric bodies, find an analogy in man himself. The physical body of the planetary system is the totality of the planets. The etheric body is that which, raying out from the fixed star, streams through the system; but in physical man, it is not the case that he carries his corpse about with him, as the planetary system does. On the other hand, it has an arrangement for getting rid of the evil astral matter by means of its comets."

"The sphere of the comets extends to the sphere of the Cherubim. We must therefore first know what is the nature of the Cherubim and Seraphim in order to understand the substance and movement of the comets. The evolution of the comets is therefore dependent upon the beings of the higher hierarchies down to the Cherubim; the evolution of the fixed star is dependent on the beings of the higher hierarchies down to the Spirits of Wisdom; the evolution of the planet itself, apart from the human being who dwells upon it, is dependent upon the forces proceeding from the higher hierarchies down to the Spirits of Form; and that which works on the Moon is dependent on the forces which proceed from the higher hierarchies down to the Archangels."

"If we begin by describing the microcosmic man from the physical body upwards, as the being standing immediately before us, we must, in the planetary system, begin by describing the corpse and must also find in the physical system the substances of the cometary bodies, which are the external expression for the *purifying astral thunderstorms* in the planetary system."

Lecture IX

"The Regents, as it were, of the principal group-forms of the animals live on our planets; they had to withdraw into these planets in order to work down with their forces at the right distance from the Earth and from the right direction. For only from these directions in space can come that which builds up the principal animal forms in the right way. Now, if the planets were only to allow these forces to stream down upon our Earth, we should not actually have the multiplicity in the animal kingdom we now have, we should only have seven principal forms. Once upon a time, in far distant ages, there were only the seven principal animal forms; but these seven forms were very mobile, determinable, so soft and plastic in their formation that they could easily be transformed; one special form into another and again other forms into others; this actually occurred at a later period of time."

"All these forces have worked together to differentiate the seven original groups of the animal kingdom; so that the whole multiplicity of our animal forms on the Earth arose from the fact that the forces of the planets are actually the abode of the group-souls, the group-egos of the animals; and that these beings fulfil their tasks from these centers, for only from there can they do so."

"It is then seen that these group-egos of the animals are the offspring of that category of spiritual beings to which I have referred in this course of lectures as the Spirits of Motion. Thus, we must look upon the group-souls of the animals as the offspring of the Spirits of Motion."

"These group-egos of the plants are the offspring of the Spirits of Wisdom, just as the group-egos of the animals are the offspring of the Spirits of Motion. Thus, in the group-egos of the plants, we have to recognize the offspring of the Spirits of Wisdom."

"We have now reached a position in which we can allude to the functions of a certain category of such Spirits of the Rotation of Time. At this point, we can state that certain Spirits of the Rotation of Time unite the forces of movement coming down from the planets to the plants that work spirally with the forces which come down from the Sun. Both these forces are brought together at a definite time by the Spirits of the Rotation of Time, and indeed at that time of the year when the plant progresses towards its fructification."

"To occult vision, the ring of Saturn is absolutely the same phenomenon as the comet's tail. If you were to take the ring of Saturn as it circles round Saturn and open it out, you would have a comet's tail."

"In this way, it is possible to look back to the streaming in of the group-souls of the minerals into our planetary system; and again, the Signs of the Zodiac in general give us their individual positions. *It is to be noted that the two outermost planets now reckoned as belonging to our system by physical astronomy—Uranus and Neptune—did not originally belong to our Solar System; they came much later into the sphere of attraction of our system*: they then joined company and remained within it. They cannot therefore be reckoned in the same sense as the other planets as belonging to our system from [Old] Saturn onwards, for they, so to speak, belonged to it from the beginning. Thus, when we consider Saturn, especially in his ancient form, we see in him a planet which, by *sending forth etheric currents* from his own center to our Earth, *creates*—we can even say—*the substance of lead*. At the same time, we see how the group-souls of the minerals stream in; we see how these group-souls are affected when a power of attraction is exercised on them from the Sun, from which the astral body of the mineral streams out. *From the Sun, the astral body of the mineral streams out into space, from outside in cosmic space, the ego of the mineral streams in*. When these currents are united, something takes place which, in a modified way, expresses itself, as it were, in a fructification of the group-ego by the astral body, and by this means alone does the mineral come to its perfection. Now, if we go back to the comet, here, too, we have something which, in fact, streams in from cosmic space: a similar stream of beings to the group-souls of the minerals. The group-souls of the minerals belong to the sphere of the Spirits of Willpower; but above them lie the beings who essentially form the basis of cometary life."

"But as everywhere there are luciferic Beings, so also within the comet there are such as stand at the stage of the Thrones, not of the Cherubim and Seraphim. That is why the comet acquires a mineral nature; appears as a mineral intervention in the planetary system; in other words, we have to look upon the comets as cosmic bodies which fly in from the cosmos after the planetary system is already formed, and thus, do not come as far as the bodies composing the system itself, but remain behind at a considerably earlier stage."

"When at night the starry heaven is illuminated, Phosphorus actually works down upon us from countless points: and everywhere in the universe, we find the possibility of formation only through the cooperation of the opposing forces; through the combined working of the normal spirits of the hierarchies with those who are rebels—that is, those who have remained behind. Unillumined to physical eyes but visible to spiritual sight, is the starry world through the normal Spirits of Wisdom; it became luminous to physical eyes, it is revealed in Maya through Lucifer or the luciferic spirits who are, and must be, active everywhere."

"The group-souls of the minerals are to be found in the sphere of the Thrones, those of the plants in the sphere of the Spirits of Wisdom, and the animals in the sphere of the Spirits of Motion; but man has so received his group-soul that with the inflowing of his ego, a group-soul was originally given him, as an emanation from the Spirits of

Form. This group-soul of man was originally allotted by the Spirits of Form to be a unitary soul for the whole of humanity. What differentiated this group-soul into such variety that differences of race, differences of tribe arose? This was brought about through the action of other spirits. Man was created to be 'One' all the world over; in this unity, the primeval ego of man was to assert itself as a group-soul dwelling in all men, a group-soul which had descended to the physical plane."

"Just as the external form only of the minerals can be brought into being by the Spirits of Form, so by these same *Spirits of Form was the group-ego created for humanity*, which was then differentiated by the activity of other beings of the various hierarchies."

"As we have seen, all the planets, all the heavenly bodies stand in connection with beings of the higher hierarchies, and so, too, is it with the Moon. The Spirits of Wisdom founded a colony upon the Moon in order to preserve the equilibrium; and so, from the direction of the Moon, there also work in upon humanity compensating spirits against Lucifer, who approached man as a tempter; and just as he disseminated light, so, too, did his spiritual principle sink down into the human soul."

"Just as in space sunlight is reflected from the Moon, so Jehovah is revealed as a reflection of the principle of the great Sun-Spirit whom you may designate by whatever name you will—Vishvakarma, as the ancient Indians called him; Ahura Mazdao, as He was called by Zarathustra, Osiris by the ancient Egyptians, or as the Christ, as He is known to the Fourth Post-Atlantean Period of civilization—that is, the esoteric comprehension of Jehovah. He is Christ reflected by the moon-principle..."

"With regard to the heavenly bodies which, as the expression of the wonders of the universe, shine forth from space, that is in many respects, very difficult; because in the heavenly bodies there is a complicated cooperation of the beings of the various hierarchies, and because everything which takes place in cosmic space can only be comprehended if, behind all matter, even behind the substance of light itself, we find the Spirit or Spirits. Behind all this Spiritual Life lies the Universal, Divine Fatherhood, an Omnipresent and ever-working All-Divine Life, which, before it comes to expression in the physical, is differentiated into countless worlds of Spiritual Hierarchies. We look up to these worlds, however, and see within them, that which works down into our kingdoms of nature, and is the foundation of all the wonders of the heavens. For even in our kingdoms of nature, either the hierarchies themselves are revealed or their offspring. When we thus look out into the spaces of heaven, we can, through such reflections, also gain a moral impression which must, if we allow the mighty operations of the hierarchies in cosmic space to gain a little influence over us, result in our being drawn away from the passions, desires, impulses, and concepts which our physical earth-life brings to maturity."

"When men seek their knowledge there, where the vision of all the humanity of the Earth can be directed, and where the knowledge common to all can be obtained—knowledge which unites and does not separate—when men actually reach that Heavenly Language which expresses the significance of the various religious founders and inspirers of humanity, then will the anthroposophical ideal of a tolerant and unbiased consideration of all religions and cosmic conceptions be really able to appear. Men will no longer quarrel when they no longer claim for their own group a particular bearer of religion or stream of civilization but seek for the origin of these bearers outside in cosmic space. In this sense, such a contemplation may acquire great moral importance if, in much which formerly brought divisions and disharmonies upon earth, peace and harmony are established. First, we must learn to read the mighty writing given us in the forms and movements of the heavenly bodies—learn to read how, in reality, not different but the same spirits, work for each single individual on Earth—that they belong to all men."

Rudolf Steiner on the Nature of Electricity

"Light decays and the decaying light is electricity. What we know as electricity is light that destroys itself within matter." GA 84

"Electricity is light in the sub-material state. Light is there compressed to the utmost degree." GA 92

"Electricity in nature is, at the same time, a moral element. When we enter the sphere of electricity, we penetrate simultaneously into a moral sphere. Electricity [man-made] contains, to be sure, moral impulses, impulses of Nature, but these impulses are immoral; they are instincts of evil, which must be overcome by the higher world." GA 220

"The greatest contrast to electricity is light. If we look upon light as electricity, we confuse good and evil. If we contemplate electricity today, we contemplate the images of a past moral reality that have turned into something evil." GA 220

"We cannot shut our eyes to the fact that we must live with Ahriman. But we must live with him in the right way, that is to say, we must not allow him to have the upper hand." GA 220

"When even light is conceived of electrically, then the qualities of Evil are attributed to the divinity of Good. It is really terrifying to see to what a great extent the modern contemplation of Nature has unawares become a demonology, a worship of demons!" GA 220

"Radiant [wireless] electricity will presently affect them in such a way that they will no longer be able to understand the news which they receive so rapidly. The effect is to damp down their intelligence." GA 327

"Electricity is not intended to work into the realm of the living—it is not meant to help living things especially; it cannot do so. You must know that electricity is at a lower level than that of living things. Whatever is alive—the higher it is, the more it will tend to ward off electricity. It is a definite repulsion." GA 327

"People will master the forces of nature to a high degree, as we have seen with wireless telegraphy and aeronautics. It is not without consequences whether the air is filled with spiritual thoughts or with thoughts of material needs. This will engulf our entire planet." GA 104a

"Seen in their true form, it can be said that these electromagnetic waves will work back into the forces of the Earth during a certain age. Then, according to good and evil, earthquakes and earth tremors will appear as the effects of human deeds." GA 104a

Steiner's Celestial Realms

The world of stars has shown itself to us as a final revelation of spiritual beings who have more and more withdrawn from active participation in the physical domain. But the connection is coming alive again due to Christ's enlivening of the etheric realm that surrounds the Earth. Even more than that, the Earth, Moon, Sun, and Stars react to what humans 'send out into space.' Human light may not be bright, but it shines, nonetheless. As Steiner says in *An Outline of Esoteric Science*: "Spiritual-psychic causes produce in the celestial bodies positions and motions through which spiritual conditions are manifested on the physical plane." Humans affect the cycles and emanations of the Sun, and thus the climate of the Earth. Mankind and the Cosmos live in a symbiotic relationship; both need each other to exist in a healthy fashion. The supposedly dying, entropic universe is actually alive and reacting to human consciousness.

Rudolf Steiner gave the Isis-Sophia Meditation at Christmas 1920, which reflects the concern for the apparently dying cosmos.

Isis Sophia
Wisdom of God,
Lucifer has slain her
And upon the wings of the forces of the world
Has carried her away to distant realms.

The will of Christ
Working in the human being,
Will rescue from Lucifer
And on the boats of spiritual knowledge
Awaken in human souls
Isis-Sophia,
The wisdom of God.

For Rudolf Steiner, the celestial bodies are the homes of spiritual hierarchies. In the First Rank of the Hierarchy there are some of the most sublime and lofty of all the beings of the Cosmos, the Thrones, Cherubim, and Seraphim, as they have been called in esoteric traditions. It is the Thrones, or Spirits of [Divine] Willpower, who create the movements of the planets through space, while the Cherubim regulate the different movements and bring about harmony between the planets. They are the messengers of the solar system, as angels are messengers for the human world. The Seraphim, Spirits of Love, have a similar task with respect to the entire world of the stars. They bring about a mutual understanding between our Sun and other stars, between fixed star and fixed star, bringing the whole Universe into balance and order.

Seraphim	Comets
Cherubim	Comets
Thrones	Fixed Stars
Spirits of Wisdom	Fixed Stars
Spirits of Movement	Planets
Spirits of Form	Planets
Archai	Moon
Archangels	Moon
Angels	Earth
Human Being	Earth

The Cosmos is filled with spiritual beings, as indicated in the chart above. Understanding the nine ranks of the spiritual hierarchies creates a cosmogenesis that leads to a complete anthropogenesis. Both are necessary to understand how 'gifts' and 'donations' of spiritual beings in the far past are still going through metamorphosis and transformation in our current world. Minerals, plants, and animals arose in the past and have transformed through heterochronic metamorphosis that laid the foundations of the physical world. The stars, planets, and our Sun once spoke and told their secrets, but today they are silent to the modern astronomer. Perhaps someday, the stars will speak to humans again as they did in the past. Rudolf Steiner gave a verse to Marie Steiner to highlight that reality on Christmas Day, 1922:

The Stars once spoke to Man

It is World destiny that they are silent now.

To be aware of the silence

Can become pain for earthly Man.

But in the deepening silence
There grows and ripens
What Man speaks to the Stars
To be aware of the speaking
Can be strength for Spirit-Man.

Below, we present many profound indications of Rudolf Steiner concerning astronomy, light, and the spiritual hierarchies. Each quotation demonstrates his comprehensive understanding of natural science that was light-years ahead of modern science.

***Third Scientific Course, Astronomy*, Rudolf Steiner, January 1-18, 1921, GA 323**

Lecture I

"He alone is studying reality, who on the one hand studies the starry skies and on the other hand the development of the human embryo."

"Every cell in its spherical form is no less than an image of the form of the whole universe. And so, you have in the whole formation of the framework of the cell a reflection of the relationships existing in the planetary system—altogether in the whole starry system."

Lecture II

"It is impossible to understand the phenomena of cell-development, especially of the sex-cells, without calling to our aid the realities of astronomy, which lie apparently so far removed from embryology."

"In considering a thing mathematically, we lift the observation out of the sphere of external reality."

"The head, the part of the human form which in its outer configuration copies universal space,—namely the sphere, the spherical shape of the Universe as a whole—the head is exposed by life in polar regions to what comes from the Cosmos outside the Earth. In the tropics, the metabolic system in its connection with the limbs is exposed to earth-life as such."

"In polar regions, the sun-light represses man's inner development. In the tropics, what shoots up from the Earth represses his inner powers."

"Man is so placed in the Universe as to be more coordinated with the cosmic surroundings of the Earth in his head, his nerve-senses system, and with the earth-life in his metabolic system. And in the temperate zones we shall have to look for a kind of perpetual harmonization between the head-system and the metabolic system. In the temperate zones, there is a primary development of the rhythmic system in man."

"The head-system is more related to the whole Cosmos, the rhythmic system is the balance between the Cosmos and the earthly world, and the metabolic system is related to the Earth itself."

"Lunar life is just as much connected with the rhythmic system of man as is the solar life with his nerves-and-senses system."

Lecture III

"Inasmuch as man individualizes himself from the Cosmos, he develops the faculty of memory; inasmuch as he still lives within the Cosmos, developing more his subconscious functions, something in the nature of a common experience with the Cosmos arises, connected with the Moon-processes in the Cosmos. This experience remains, just as a past experience remains in our memory, and later it emerges in an inner constitutional process, like a remembrance which has been drawn into the body and has become organic."

"Now take the following. If we consider the daily influence of the Sun, we find an inner activity of soul and spirit; if we consider the yearly influence of the Sun, then we find laws of growth belonging to the outer physical body."

Thus, we can say, for the Sun life:

Soul and Spirit: Day

Physical bodily nature: Year

Thus, for the Lunar life:

Soul and Spirit: 28 days of activity

Physical bodily nature: 28 year's activity

"Let us now seek the corresponding phenomena for the physical body. What will they be? You can find it for yourselves by deduction. We will have bodily and physical effects with a 28-year period. As a day here corresponds to a year, we shall have 28 years."

"You need only remember that 28 years is the period bringing us to our full inner maturity of growth. It is then that we first cease to be in the ascending scale of growth. Just as the Sun works upon us from outside in its yearly activity, in order to complete in us an outward process corresponding to the daily process in the inner life of soul and spirit, so something works in the Cosmos in a 28-year period, organizing us from outside even as the female human being is organized inwardly."

"As the daily Sun-life is related to the yearly Sun-life in regard to man, so the 28-day Moon-life is related to the 28-year Moon-life with respect to the whole man (the former belonging, in effect, more to the human head)."

Lecture IV

"Thus, in effect, in employing mathematics in astronomy, we come at one point into chaos. And by pure observation in embryology, we have at a certain point nothing before us but chaos; it all seems chaotic at first, observation is impossible. Then we come out of chaos into the realm of geometry."

"Now therefore, we can say: When in embryology we try to follow up the real process by observation, we emerge out of a sphere which lies about as near to our understanding as that which is beyond the incommensurable numbers. In astronomy, on the one hand, we proceed with our understanding up to the point where we can no longer follow mathematically. In embryology, on the other hand, our understanding begins at a certain point, where we are first able to set to work with something resembling geometry."

"I shall be wrong if I take mathematics as constituting reality, for then in certain fields, reality will simply contradict me. Such a contradiction is the one I spoke of, where incommensurability enters in, in the study of celestial phenomena."

Lecture V

"What confronts us in embryology emerges as if from indefinite, chaotic regions, and from a certain point onward can be grasped in picture-form, or even geometrically."

"Astronomy will not go into the human head; it simply will not fit there."

"Through his head, he takes part in the astronomical, through his metabolism in the meteorological domain."

"Think what this means. All that goes on in the ovum before fertilization is, so to speak, within the domain of the human being's own inner organic process. But what happens after fertilization and is brought about thereby—the human being opens himself to the Cosmos. Cosmic influences here prevail."

"Astronomy, practiced as hitherto, will never lead to a grasp of reality; it will only give us pictures. And embryology, though in this realm we seize reality, will not enable us to penetrate reality with ideas and mental pictures. Astronomical pictures of the world are poor in reality; embryological pictures are poor in idea—we fail to penetrate the facts with clear ideas."

Lecture VI

"Looking into the future, there will be another Ice Age in these regions when the vernal point reaches the sign of Scorpio. This rhythm is contained within what takes its course in 25,920 years. Although admittedly of vast extent, it is a true rhythm, none the less."

Lecture VII

"We need to advance from the merely quantitative to the qualitative aspect."

"The female human body—to the environment is, in a metamorphosed form, the same as the relation of the eye to the environment."

"Into the material domain, the female life is the life of fantasy or fancy of the Universe, whereas the male is that which forms the contours and sharp outlines. It is the male which transforms the undetermined life of fancy into a life of determined form and outline. Seen in the way we have described in today's lecture; the process of sight is none other than a direct metamorphosis of that of fertilization; and vice-versa."

"The clear differentiation between the life of the senses and the organic life of the human being as a whole had not yet taken place in the time preceding the last Ice-Age. The human being still enjoyed a more synthetic, more 'single' organic life. Since the last Ice-Age, man's organic life has undergone, as one might say, a very real 'analysis.' This too is an indication that the relation of the Earth to the Sun was different before the last Ice-Age from what it afterwards became. This is the kind of premise from which we have to take our start, so as to reach genuine pictures and ideas about the Universe in its relation to the Earth and man."

Lecture VIII

"The ratios between the periods of revolution of several planets are incommensurable numbers,—incommensurable magnitudes. By calculation, therefore, we do not reach the innermost structure of the celestial phenomena. Sooner or later, we must leave off. What is alive in the planetary system is precisely what we cannot calculate."

"In the formulae of physics, wherever the force of pressure occurs as regards ponderable matter, we have to replace it by a force of suction as regards the ether. In other words, if we insert a plus sign for the intensity of a force in the realm of ponderable matter, we must give a minus sign to the corresponding intensity in the ether."

"Whatever in our formulae is to represent the effects of light, must bear a negative sign. For air or gas, the sign must be positive. For warmth, on the other hand, the positive and negative will have to alternate."

"For the planetary system, we also have to pass from the positive—that is, gravitation—to the corresponding negative, the repelling force."

"There is the planetary system; it receives something new into itself, namely the effects of a comet. There is the ovum; it receives into itself the fertilizing effect of the male cell, the spermatozoid."

Lecture X

"We can think of the whole celestial sphere as being mirrored in the Earth: We conceive the mineral kingdom of the Earth as an outcome of this reflection, and we conceive that what lives within us, enabling us to understand the mineral kingdom, comes from what surrounds us out in celestial space."

"Therefore, we must distinguish, in the genesis of the Earth as well, between two phases: One in which active forces work in such a way that the real Earth itself is created, and then a later phase of evolution, in which the forces work so as to create the human faculty for understanding the realities of the Earth."

Lecture XI

"As we have seen, if we wish to relate Man to the Cosmos, we must assign the metabolic system to what is earthly,—what comes to man in a radial direction. Whereas we must assign the forming of the head to all that derives from the Great Sphere,—that sends its lines of influence, as it were, from the celestial sphere towards the center of the Earth, even as the radius reaches outward with its lines of influence to its surroundings. We saw this in the construction of the typical long bones or tubular bones by contrast to the skull bones, the latter being sphere-like, or like a sector of a sphere."

"Envisaging this difference, we must relate it, to begin with, to what appears to us in the relation of the Earth to the Celestial Sphere."

"Apply this principle—the inner mobility, as it were, of movement in itself—to Nature. Try to express in equations, this that 'moves the moving.' You will then find it possible, mathematically, to penetrate what is organic. You will come to say, for it can well be formulated thus: The axioms of rigid space—space immobile in itself—lead to an understanding of inorganic Nature. Conceive of a space that is inherently mobile—or algebraic equations whose very functionality is in itself a function—and you will find the transition to a mathematical understanding of organic Nature. This, incidentally, is the method which should accompany the efforts now being made to investigate the transition-forms from inorganic Nature to organic, as regards shape and form at least."

"Take now the presence of the loop-making tendency in the human body and compare it with what confronts us, admittedly in a more irrational form, in the forms of movement of the planets. You will then realize: The 'apparent movements' of the planets, as we are wont to call them, in a most striking way inscribe, in forms of movement in the heavens, what in the human body is a form as such—a characteristic, fundamental figure. Therefore, to say the least, we must in some way correlate this basic form in the human body and these phenomena in the Heavens. And we shall now be able to say: Behold the loop. It always appears when the planet is relatively near the Earth,—therefore when we, being on the Earth, are in a special relation to the planet."

"We, in our process of formation, are thus related to the planet: We are exposed at one time to the planet's loop and at another to the remainder of its path. We can therefore relate, what lies nearer to our head, to the loop, and what belongs more to the remainder of our body, to the planetary path outside the loop."

"If we develop a feeling for morphology in the higher sense, we can do no other than assign the human form and figure to the planetary system. The planets must somehow correspond to the whole man; the fixed stars only to what forms the head of man."

"For, as we move with the Earth (we must project this backward, once more, backward in time to the embryo-period of our life),—as we move with the Earth, what we have in us comes into being, formed as indeed it is by the very forces of movement. In the movements we see up yonder in their seeming forms and pictures, we have to recognize the cosmic movements we ourselves are making in the year's course. We realize it as we approach the true aspect of the loop-curve."

"The planetary movements are, to begin with, images—images of—movements you yourself accomplish with the Earth in the year's course. (We shall see in time, how a synthesis arises from the different planetary movements.)"

"If then we see the human being as a whole and his projection to the Cosmos, we are led to recognize that the true form of movement of the Earth in the year's course will be the loop-curve or Lemniscate."

"The path of the Earth itself is a loop-curve—quite apart now from its relation to the Sun or any other factor. What is projected then, for our perception, the planetary paths with the loops they make,—we must regard as the projection by the planets of the loop-path of the Earth on to the vault of heaven, if we may formulate thus simply a very complicated set of facts. As to why, when the planet draws near the loop, we have to leave the rest of the path open during a relatively short space of time,—the reason lies in the fact that under certain conditions the projection of a closed curve may appear open."

Lecture XII

"Yesterday I showed that wheresoever we may look in the human body, we shall find the formative principle of the looped curve or lemniscate, save for the two outermost polarities—the radius and the sphere. Thus, in the human body, we perceive three formative principles: The sphere, with its activity primarily going inward, the radius, and between these the looped curve or lemniscate."

"Thus, to sum up, the building principle of the human form, howsoever modified, is based on this; the metamorphosis emerges between head and limbs—the limbs with the metabolism which belongs to them—and in the great Universe, this answers to the contrast between those planets that form them in opposition to the Sun. Between the two is then the Sun itself."

"We therefore have to imagine a certain contrast between the paths of the superior and of the inferior planets; and in the Sun's path, a quality midway between the two."

"Sun and Moon on the one hand and of the loop-forming planetary paths on the other. The planetary paths with their characteristic loops quite evidently correspond to what makes vortices and vertebrae,—to what is lemniscatory in man."

"Look simply at the human form and figure and think of its relation to the Earth; we can do no other than connect what is radial in human form and stature with the path of the Sun, even as we connect what is lemniscatory in form with the typical planetary path."

"You see then what emerges when we are able to relate to the starry Heavens the entire human being, not only the human organ of cognition. This, in effect, emerges: In the vertical axis of man, we must in some way seek what answers to the Sun's path, whilst in all that is lemniscatory in arrangement we have to seek what answers to the planetary paths,—lemniscatory as they are too, though in a variable form. Important truths will follow from this, we must conceive, once more, that through his vertical axis, man is related to the Sun's path."

"The Earth is following the Sun! The Sun speeds on ahead, the Earth comes after. This then must be the essence of the matter: the earthly and the solar orbit in some way coincide, and the Earth somehow follows the Sun, making it possible as the Earth rotates for the Earth's radii to fall into the solar path, or at the very least to be in a certain relation to it."

"As you are well aware, to explain all the phenomena, astronomy today must have recourse not only to the primary notion of a stationary Sun supposed to be at the focus of an ellipse along which the Earth is moving—but to a further movement, a movement of the Sun itself towards a certain constellation. If you imagine the direction of this movement and other relevant factors, then from the several movements of Sun and Earth, you may well be able to deduce a resultant path for the Earth, no longer coincident with the ellipse in which the Earth is said to be going round the Sun, but of a different form which need not be at all like the supposed ellipse."

Lecture XIII

"In effect, the Sun reveals his influence comparatively simply through the heart. Jupiter, we must admit, spins like a wheel in our head,—it is the whirling epicycle. Whilst in a different sense, Venus goes through beneath our heart."

"We have put two things, and of different kind, before us: on the one hand the paths of the epicycle-centers and of the points on the epicyclic circumference, revealing a quite different relation to the Sun for the superior and inferior planets respectively; on the other hand, the prolongation of the plant-forming process speeding on into the mineral, whilst the animal-forming process turns back upon itself to become man."

Lecture XIV

"We are always living in the Moon, so that what truly deserves the name of 'Moon' is, in reality, a realm of force, a complex of forces that pervades us all the time."

"We must realize that what is visible in the heavens is no more than a fragmentary manifestation of cosmic space, which in reality is ever filled with substance. Develop this idea: you live immersed in substance—substances manifold, inter-related. Then you will get a feeling of how very real a thing it is. The accepted astronomical outlook of our time has replaced this 'real' by something merely thought-out, namely by 'gravitation' as we call it. We only think there is a mutual force of attraction between what we imagine to be the body of the Moon and the body of the Earth respectively."

"The Moon pervades the Earth, including the Earth's hydrosphere. In the Moon's sphere, something is going on which at one place it manifests in a phenomenon of light. We need not think of any extra force of attraction. All we need think is that this Moon-sphere, permeating the Earth, is one with it, one organism all together, an organic whole."

"Think then of this: Here we are in the Earth-sphere, in the Sun-sphere, in the Moon-sphere, and in others too. You never gain a clear notion of it if you conceive the heavenly bodies as being far away yonder in space."

"If we somehow succeeded in finding an ideal point-of-intersection of the effects of Sun and Moon and Earth, if we could ascertain the movement of this point for every individual human being, it would lead us far indeed towards an understanding of what we may, perhaps, describe as movement, speaking of Sun and Moon and Earth."

Lecture XV

"We may not therefore take for granted that we shall ever be able to explain the cosmic phenomena on the accustomed basis of our geometry, that is to say, within a rigid three-dimensional space."

Lecture XVI

"That which is commonly thought of as the physical constitution of the Sun can never really be understood with the ideas we derive from earthly life."

Lecture XVII

"As the plant grows away from the Earth, so have we to imagine this super-physical man-plant growing in from cosmic space, even from the Sun, towards the center of the Earth."

"Earth and Sun are moving, following one another, along a lemniscatory spiral..."

"It is as though, with respect to the planets, Earth and Sun were taking turns, alternately being in the center of the system."

"Tracing the path of Earth and Sun (now, once again, apart from the planetary system as a whole) we have then to imagine a Lemniscate in which the Earth is following the Sun. Incidentally, you may also see in this a possibility of giving meaning to the idea of gravitation. The one draws the other after it: that is the underlying principle. Think of it in this way, and you will no longer need the somewhat questionable quality of gravitational and tangential forces, for they are here reduced to a single force. Think it through thoroughly and you will find it so. You must admit, it is a rather problematical feature in the Newtonian conception. We are to think of the Sun in the center and the planets around it—endowed, one and all, with a kind of "shove" in the tangential direction, one and all, without presupposing which the Newtonian system would break down."

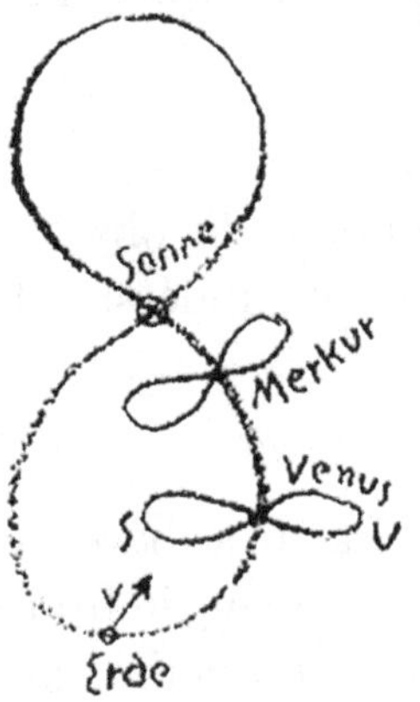

Figure 6

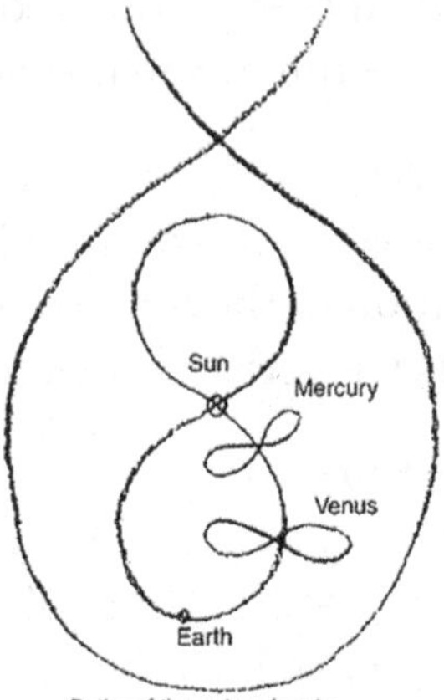

Figure 7

"Taking this then to be the path of Earth and Sun,—if you wish to bring out in perspective, along with the course of Earth and Sun, the path-forms of the other planets, you must imagine the paths of the inferior planets somewhat in this way (small lemniscates in Figure 6)."

"On the other hand, taking this once more to be the path of Earth and Sun and this the path of the inferior planets, you must imagine the corresponding paths of the superior planets to be lemniscates like this (Fig. 7). I should now have to go on drawing upward, but the nearest part would be like this. And now this lemniscate 2 moves on, makes its way through,—through the lemniscate of the superior planets."

"It is a system of lemniscates in determined order and relation. Such are the paths of the planets; such also is the path of Earth and Sun. Now you will easily harmonize what I have here presented in the diagrammatic outline, with the fact that we see the loops of Venus and Mercury in conjunction and those of Jupiter, Mars, and Saturn in opposition. In our perspective, it is the necessary outcome."

"What we behold in Venus and Mercury is far more akin to what is living as a cosmical reality in us ourselves. Whilst what we see in the paths of the superior planets is more akin to the fixed-star heavens in general."

"An inferior planet has a path, making a lemniscate loop-curve the center of which is the Earth-and-Sun path itself. A superior planet, on the other hand, embraces the Earth-and-Sun path in its own lemniscate-loop. Such is the essence of the matter; the thing itself is so complicated that the mental pictures we can form scarcely be more than diagrammatic."

Lecture XVIII

"The Sun must be conceived as a hollowing-out, shall we say, of cosmic matter, a hollow space, a hollow sphere,—a sphere enveloped by matter,—in contrast to the Earth where we have denser matter enveloped by more attenuated. As to the Earth, we think of air around it. Air is outside and denser matter inside. For the Sun, it is the opposite: as we go inward, we go from relatively denser matter into more attenuated and at long last into the very negation of matter. The Sun is not only a more attenuated heavenly body, of a materiality less dense than earthly matter, but if we call the Earth's materiality positive, then in the Sun—in the Sun's interior—we shall have negative matter, in a certain sense. We only do justice to the phenomena if we conceive that there is negative matter in the inner space of the Sun."

"Now, my dear friends, as compared with positive matter, negative matter is suctional. Positive matter exerts pressure and negative suction. And if you now conceive of the Sun as a collection of suctional force, you need no further explanation of gravitation. This is the explanation. Now think of it as I explained yesterday. The movement of Earth and Sun is such that the Earth follows the Sun in the same path, in the same direction. Here then you have the cosmic relation between the Sun and Earth. The Sun as a gathering of suctional forces goes on in front, and by this suctional force, the Earth is drawn on after, moving through cosmic space in the same course and in the same direction in which the Sun thrusts forward."

"This force of suction therefore, proceeding from the Sun, works also in the human being, permeating his etheric body from above downward. Upon the human being—the human body in this instance—two opposite entities are at work; Sun-entity, Earth-entity."

"Here you already have the opposition—the working of the ponderable and of the imponderable force. It is the interplay of the ponderable pressure from without inward and of the imponderable from within outward which gives you the conscious sensation of pressure."

"If in our mind, we see all these things clearly and comprehensively, we may truly say that the polarity of Sun and Earth into the midst of which the human being is placed, is felt by us in every sense-perception. In like manner, everything about the human being can be traced in such a way as to perceive the cosmic realities that are involved. Cosmic forces work into the human being upon every hand."

"In effect, my dear friends, if I were to trace the paths of the heavenly bodies, I should really have to go out into the Universe and trace them ever anew, varying them all the time."

"Comets are ever-nascent phenomena, perpetually coming into being, passing away again. Out there—outside the system—we need not imagine it to exist at all. It is not there to begin with, but becomes; then, at the perihelion, changes the gesture of its form, which is in fact ever-becoming, ever-nascent. Out there, at last, it melts away again and is no more. The comet comes into being and passes away; that is its very nature."

"Altogether, we must look upon the comet as a fleeting thing. In relation to Sun and Earth, it is a phenomenon of compensation between ponderable and imponderable matter,—a meeting of the two kinds of matter, which do not immediately balance out, as when light extends in air."

The Light Course, Rudolf Steiner, Lecture I, December 23, 1919, Stuttgart, GA 320

"You see then how the diverse ways in which the life of man unites with the material element that underlies it bring about, upon the one hand, the *submersion of the Will in matter* and on the other hand the *lightening of Will into Intelligence*."

"In the human being, we then see upon the one hand the lightening into Intelligence, brought about by one kind of connection with the material life—connection namely with the buoyancy which is at work there. Whilst on the other hand, where he has to let his Will be absorbed, sucked up as it were, by the downward pressure, we see men being put to sleep. For the Will works in the sense of this downward pressure. Hence, our intelligence is to some extent permeated by Will. In the main, however, *what is at work in intelligence is the very opposite of ponderable matter.* We always tend to go up and out beyond our head when we are thinking."

"We, with our etheric body, swim in the light-ether. The light remains invisible to us; we swim in it. Here it is as with space and time; we ought not to call them objective, for we ourselves are swimming in them. So too, we should regard the light as an element common to us and to the things outside us; whilst in the colors, we have to recognize something that can only make its appearance inasmuch as we, through our astral body, come into relation to what the light is doing there."

"Then in the last resort, you will be bound to recognize that the distinction is: we share in the warmth-conditions of our environment with our physical body and in the light-conditions, as we said just now, with our etheric body."

"In light and sound and warmth, we ourselves are swimming, so to speak, as was described in former lectures. The same cannot be said so simply of our relation to electrical phenomena. We do not perceive electricity as a specific quality in the way we perceive light, for instance. Even when electricity is at last obliged to reveal itself, we perceive it by means of a phenomenon of light. This led to people's saying, what they have kept repeating: "There is no sense-organ for electricity in man." The light has built for itself in man the eye—a sense-organ with which to see it. So has the sound, the ear. For warmth too, a kind of warmth-organ is built into man. For electricity, they say, there is nothing analogous. We perceive electricity indirectly."

"Our experience of light, sound, and warmth belongs—to a high degree at least, if not entirely—to the field which we comprise and comprehend with our sensory and thinking life. Above all, this is true of the phenomena of light. An open-minded study of the human being shows that all these things are akin to our conscious faculties of soul.

On the other hand, the moment we go on to the essential qualities of mass and matter, we are approaching what is akin to those forces which develop in us when we are sleeping. And we are going in precisely the same direction when we descend from the realm of light and sound and warmth into the realm of the electrical phenomena."

"We have no direct experience of the phenomena of our own Will; all we are able to experience in consciousness is our thoughts about them. Likewise, we have no direct experience of the electrical phenomena of Nature. We only experience what they deliver, what they send upward, to speak, into the realms of light and sound and warmth, etc. For we are here crossing the same boundary as to the outer world, which we are crossing in ourselves when we descend from our thinking and idea-forming, conscious life into our life of Will. All that is light, and sound, and warmth, is then akin to our conscious life, while all that goes on in the realms of electricity and magnetism is akin—intimately akin—to our unconscious life of Will."

"When in Nature we pass on from the phenomena of light, sound, and warmth to those of electricity, we are descending—in the realm of nature—into phenomena which are related to the former ones as is the Will in us to the life of Thought."

"The phenomena of sound and tone and light are akin to the conscious element of Thought and Ideation in ourselves, while those of electricity and magnetism are akin to the sub-conscious element of willpower. Warmth is between the two. Even as Feeling is intermediate between Thought and Willpower, so is the outer warmth in Nature intermediate between light and sound on the one hand, electricity, and magnetism on the other."

"As in the spiritual realm, we differentiate between the luciferic, that is akin to the quality of light, and the ahrimanic, akin to electricity and magnetism, so also must we understand the structure of the phenomena of Nature. Between the two lies what we meet with in the phenomena of Warmth."

Universe, Earth, and Man, Rudolf Steiner, August 4—August 16, 1908, GA 105

"Let us now consider for a time, the evolution of our Earth. If it is observed occultly, we see it as a body inhabited by man on which he carries out his development. This development has only been made possible through the Sun and the Moon having separated from the Earth, so that its forces were held in balance between the two. At the time when the Earth was itself still Sun (if we may call it so), it passed through an evolution in which it was incorporated with the Sun. The Sun was then itself at the planetary stage of existence, and was inhabited by Archangels, but because of advancing development, it was possible for part of that which was embodied in it to rise to a higher existence at the cost of that other part which it sent forth as the earth-moon. In the great universe, evolution proceeds in such a way that things which, for a time, have progressed side by side separate; the one expanding into higher regions, the other descending into a lower state. In order that certain beings might develop high enough, the Sun had itself to become a body fit for their habitation; it advanced from planetary existence to fixed-star existence. We have to realize that a world-being like our Sun has developed occultly from a planet to a sun. A sun is a planet that has progressed."

"As was pointed out in the last lecture, after everything had united again and the Sun had once more, at a certain period, separated from the moon-plus-earth, man continued to dwell for a long period upon the Earth, on into the present earth-period, without the spiritual sun-forces. Then, through the advent of Christ, the spiritual forces of the Sun found again a place upon the Earth. Now, if the Christ is embodied in the Earth, man must become more and more mature through receiving into him the Christ-principle; the material form of a planet depends on what it evolves in the way of beings."

"Exactly in the same way as the Sun evolved to its present exalted position, by withdrawing the finer substances because the Sun-being had need of them, so also will the Earth. The substances of the Earth will have then so changed as to be suited to man, or rather to what, in the distant future, will have developed from man and from the earth-beings he bears along with him; for when man has become powerful, he will draw other earth-beings along with him. What will happen then? If man fills himself ever more and more with the Christ-principle, if he absorbs more and more of the sun-forces which descended to Earth with the Christ, he will himself grow ever more Christ-like and will irradiate the whole heart with the Christ-principle."

"When clairvoyant vision sweeps out into the universe and we investigate the other planets belonging to our system, we find in all those belonging to our Sun a vegetable kingdom entirely corresponding to our own—so that in our vegetable kingdom we have something that in its systematic life is a part of our whole universe."

"*Our solar system is peopled by a vegetable creation*, and were the whole matter to be considered occultly, we should see that *each planet is peopled also by its own kind of human beings*."

Elemental Beings and Human Destinies, Rudolf Steiner, Dec. 6, 1919, Dornach, GA 194

"People usually have the following trivial picture: out there somewhere is the Sun, and from the Sun, light goes out in all directions, just as from any other source of light. And you will find that wherever people follow this diffusion of light with mathematical ideas, they will say: You see, the light spreads out and out into the infinite, and then—why then does it somehow or other disappear; it gradually weakens and is lost. But this is not so. Everything that spreads out or is diffused in this way reaches a boundary, and from this boundary, it swings back again; it returns to its source in a changed form. *The sunlight does not go out into the infinite, but swings back on itself—not indeed as light, but as something else.* None the less, it does return."

"So it is, in reality, with every form of light. And so, it is with every kind of activity. All activities and influences are subject to the law of elasticity. The elasticity in them always has its boundary or limit."

"They call the velocity, usually denoted by 'v', a function of distance and time, and they arrive at the following equation: v = d/t..." But, my dear friends, that is absolutely false. The velocity is not a resultant; the velocity is an elementary principle or quality that something, be it material or spiritual, bears within it. And this velocity, we analyze; we split it up into distance and time. We abstract the two things out of it—space and time. Space and time, however, are not real things in themselves. Velocities, varying velocities, are real."

Supersensible Man, Rudolf Steiner, November 13–18, 1923, GA 231

"Saturn is not inactive in relation to the ether in which the whole planetary sphere is contained and enclosed. Seen with the eye of the spirit, Saturn rays out forces. From Saturn, radiates something that can be perceived as form. The physical planet Saturn is only one part of the picture—a part that gradually fades away before the eye of the spirit. One has the feeling that the Spirits of the World have placed Saturn there in this position in the heavens on our behalf, as it were, in order that we may have a direction in which to focus our gaze. To the eye of the spirit, it is as if someone were to make a dot on the blackboard, draw something around it, and then rub the dot out again."

"This is actually what happens in spiritual sight. Saturn is blotted out, but what is around Saturn becomes clearer and clearer and tells a marvelous story. If we have reached the point where Saturn itself is blotted out and we behold the "form" or "figure" that has been worked into the ether, we find that this form extends as far as Jupiter, where the same process is repeated. Jupiter is blotted out and what comes into being in the ether spreads out, spreads out very far; until once again a form arises in the ether, which combines with the form from Saturn to produce a picture in the heavens. We come to Mars, and the same thing happens again. Then we come to the Sun. Whereas the outer, physical Sun blinds and dazzles, we find it is not so with the spiritual Sun. All the dazzling quickly dies away when we gaze at the spiritual Sun, and a great, majestic, living picture arises from all that is inscribed into the ether—a picture that extends also to Venus, Mercury, Moon. We have, now, a complete picture with its different parts."

"Some of you may suggest that there will be occasions when Saturn, for instance, is standing at a place in the heavens where he cannot come in contact with the picture formed by Jupiter. In a wonderful way, this too is provided for. The contact is brought about in the following manner. If you were to start from a certain point lying in the East, in Asia, and draw a line right through the center of the Earth to the other side and then extend it out into the Cosmos, you would have drawn a line that is of the greatest significance for the whole field of spiritual sight. When Saturn lies outside this line, we must carry over the picture that arises from Saturn to the line; this fixes it. The pictures are fixed by means of this line. Wherever we may have found the Jupiter-picture or the Saturn-picture—and they have to be sought for—they are fixed for our sight by being brought to this line. We have thus finally one single picture. Our planetary system presents a complete picture. Do you know what this picture is? We unriddle it and discover what it is—a great cosmic picture of human skin with the sense-organs. If you take the skin of a human being, including with it the sense-organs, and try to draw the picture which corresponds to it in the heavens, it proves to be what I have just described."

"The planetary system inscribes into the cosmic ether what is present in the human being—differentiated and specialized by earthly conditions—in the spatial picture of the surface of the skin including the sense-organs. That, then, is the first thing. We discover a connection between the human being, on Earth, in respect to the form given him by the skin which encloses him, and the planetary system which shapes, forms, and builds into the ether, the archetypal, heavenly picture of earthly man."

"Now we make a second discovery. We look at the planets in movement. If we watch any particular planet, then the Ptolemaic and the Copernican systems will give us each a different picture of its course. That can very well be, the pictures of planetary movements can be interpreted in many ways. But what is far more important is that we should now be able to behold all these movements together. Suppose we are looking at Saturn, the planet that has the longest way to go and needs the longest time in which to complete its orbit. The movement of Saturn seen in conjunction with the movement of Jupiter gives a picture. Looking now at all the planets together in this way in their several movements, we have before us once again one complete picture arising this time from the movements of the planets. The picture does not tally with the astronomical descriptions of the planetary movements. Strange to say, spiritual sight does not find the pictures of ellipses which you can see drawn in astronomical maps. When we follow Saturn, for example, with spiritual sight, he reveals to us something which, in conjunction with other movements, forms itself into a figure of eight, a kind of lemniscate. Into this form enter manifold other planetary movements. So, once again, we have a picture. This picture arising from all the planetary movements reveals itself to us as the heavenly picture of what comes to expression in the human being in the nerves and the neighboring glands. The archetypal picture of the human skin and sense-organs is found by spiritual sight in the order and grouping of the planets. We have now seen what happens when we pass from this to the picture of planetary movements. If we draw an outline of the human form, we can have the feeling: This outline represents the form of the planetary system; but when we draw in the nervous system and the secreting glands, then with every stroke we are drawing a physical picture of the movements of the planetary system as they are seen with the eye of the spirit."

"We can now take another step forward in our spiritual observation of the Cosmos. Having reached the point where we obtain a picture of the movements of the planets by drawing into our outline of the human form the nerves and neighboring glands, we can go further. The several movements fade away. As we rise from Imagination to Inspiration, the movements vanish. This is of extraordinary significance. "Seeing" in the narrower sense ceases, and we begin to "hear" in the spirit. What was previously movement becoming dim and confused, until it is like a picture seen in a mist. But out of this misty picture, the Music of the Cosmos begins to form—the Cosmic Rhythms become audible for us in the spirit. And we ask ourselves: What is it we must now add to our outline of the human form, to correspond with these Cosmic Rhythms?"

"In the sphere of Art, as you know, all manner of transformations is possible. When we have drawn our outline of man and then drawn within it the nervous system, we have the feeling that we have been literally painting or drawing. But now it is not so easy to paint what we hear in the realm of Cosmic Music, for it is all rhythm and melody. If we are to represent it in our picture, we must take a brush and, following the nervous system, quickly make here a dab of red, there a dab of blue, here again red, there again blue, and so on, all along the lines of the nervous system. Then at certain places, we shall feel impelled to stop, we can go no further; we must now paint into the picture a definite 'form,' to express what we have heard in the spirit. We can indeed transform it into drawing, but if we want to place it within the contour-line, we find that at certain points we are obliged to go beyond the line and paint a new and different form, because here the rhythm blue-red, blue-red, blue-red, suddenly becomes melody. We feel we must paint in this form—and the form is what the melody sings to us! Cosmic Rhythm—Cosmic Melody. When we have completed the picture, we have before us Cosmic Music made perceptible in space, the Cosmic Music which becomes audible to the ear of the spirit when the picture of the planetary movements grows dim and disappears. And what we have now drawn into our picture is none other than the path along which the blood flows. When we come to an organ—to heart or lung, or to organs which take into themselves either something from the outside world or substances from within the body itself—at these points, we must paint a form which attaches itself in some way to the channels of the blood. Then we get heart, lungs, liver, kidneys, stomach. From the Cosmic Music, we learn how to draw these organs of secretion, and how to insert them into the blood system in our picture."

"Now we go a stage further. We pass from Inspiration to Intuition. Something new arises out of the Cosmic Music. The tones begin to blend with one another; one tone works upon another, and we begin to hear meaning in this Cosmic Music. The Cosmic Music changes into speech—Cosmic Speech that is spoken forth by the Universe. At the stage of Intuition, what was known in earlier times as the Cosmic Word becomes audible. We must now draw something else into our picture of the human being. Here we must proceed just as we proceed in ordinary everyday writing, where we express something by means of words that are formed of letters. In our picture of man, we must express the meaning of the single Cosmic Words. We find that when we give expression to these Cosmic Words and bring that expression into the drawing, we have before us a picture of the muscular and bony systems in the human being. It is just as though someone were to tell us something which we then write down. Cosmic Speech tells us something—and we draw it into the picture. In what the world beyond the Earth tells us, we have thus been able to find the human being in his totality."

"But now there is another and essentially different experience that comes to us in the course of this spiritual observation. Let us return to what was said at the beginning of the lecture about the form that is inscribed in the ether by the planetary bodies."

"We have, then, before us a spiritual, super-sensible picture of man. For in everything I have described—skin, nervous system, blood system, muscles, bones—there are, to begin with, only forces. At first, it is all a kind of picture of forces. At conception and birth, it joins with the physical embryo provided by the Earth and receives into itself earthly forces and substances. This picture—a purely spiritual but at the same time definite picture—is then filled out with earthly substances and forces."

"Man comes down to Earth as a being formed and fashioned by the heavens. He is at first wholly super-sensible, he is a super-sensible being to his very bones. Then he unites with the embryonic germ; he takes it up."

Popular Occultism, Rudolf Steiner, Lecture X, *Paths of Occult Training*, July 8, 1906, GA 94

"The future task of civilization, as a whole, consists in becoming reunited with the Spirit. Every epoch has its particular tasks. Modern science has rejected the Ptolemaic world-system as erroneous and has adopted the world-systems of Galilei and Copernicus: but for the astral plane, the Ptolemaic system is correct; for there one sets out from quite different perspectives."

"You may be astonished at the sequence in which I have placed the planets. When the Earth is here, and the Sun there, you would have thought that I should draw Mercury in the vicinity of the Sun, and Venus here. But no! For these Planets have had their names interchanged, in later astronomy. That which is called Mercury today was called Venus in ancient teachings, and that which is called Venus was called Mercury. Thus, note it well, one does not understand the ancient writings when one takes that which in them is called Venus or Mercury for the Venus or Mercury of the present day."

"That which is said about Venus has to be applied to the Mercury of today, and what is said about Mercury to Venus. For those two designations were later interchanged. On the occasion when man turned the world system topsy-turvy, when the earth was deprived of its central position, the perspective was not only changed, but the designations of Mercury and Venus were also changed.

Lightning is Plasma

Rudolf Steiner tells us that primal plasma in space helps create life, humanity, and a world of archetypal ideas. He even says that "plasma and flesh: they are basically the same substances." He also tells us that he human germ cell itself is seemingly eternal and that only half of the "germ plasma" goes over into the new body.

Plasma works with all seven ethers and can be found when light ether "thrusts down" to create electricity (static) energy in sub-nature; sound/chemical ether thrusts down to create electromagnetism, and life ether (the Harmony of the Spheres) thrusts down to create a "terrible force" [the "Third Force"].

The atmospheres of Earth are a living cycle of oxygen, lightning, and water. Water is "fertilized by warmth and light from the universe." Lightning (plasma) is all around us throughout the summer, fertilizing plants in a "flash of light." Everywhere in summer "lower lightning" (plasma) is all around us and lightning happens in us all the time because our thoughts are "lightning (plasma) and thunder flashing within us."

Rudolf Steiner tells us that lightning (plasma) is like human blood, and thunder is like human thought acting upon the nervous system. He says that "God thinks in rolling fire." Lightning (plasma) breaks through into the sensory world, radiating in from the spiritual world cleaning Earth's atmospheres. Stars (plasmas) appear to us out of the spirit like "living revelations of lightning" (plasma). Behind the beings who live in lightning flashes are the Seraphim, Cherubim, and Thrones. The Cherubim show us their tools in lightning (plasma). Lightning is the tool of the Cherubim, the fiery swords of the Cherubim.

According to Dr. Steiner, lightning derives from "universal heat" or "cosmic heat" which is the speech of the Seraphim who reveal themselves by shining. In thunder, we hear the Thrones echo lightning, the speech of the Seraphim.

Thought originates in our soul like thunder rolling through the Cosmos. When lightning flashes in the air, and when air and fire interact to produce thunder, this corresponds to the fire of our blood and the activity of our nervous system. Lightning in clouds corresponds to the fire and warmth in our blood, whereas the air outside, including all the elements it contains in the Universe, corresponds to everything that passes through our nervous system. Just as lightning in its counterplay with the elements produces thunder, so the counterplay of blood and nerves produces thought that flashes up in the soul. That is the invisible God who weaves and radiates through the Universe. Steiner informs us that "His warmth is in the lightning, His nerves are in the air, His thoughts are in the rolling thunder; and it was He who spoke to Moses in the burning bramblebush and in the lightning fire of Sinai."

Lightning comes out of overheated air, the overheated Universe, and overheated cosmic gas (plasma). As a consequence of the inner overheating of the cosmic gas (plasma), the warmth that is living in the cosmic gas (plasma) comes out. As a consequence of the greatly overheated cosmic gas (plasma), lightning flashes out. Lightning is not caused by electricity, but by the fact that the air is getting rid of, emptying away, its own heat. Steiner frankly tells us that: "It is lightning that stimulates electricity; lightning itself is not electricity."

The Cherubim are the directing powers in cloud formation, and the Seraphim are the directing powers in the lightning flash that issues from clouds. If we look upon the expulsion from Paradise as really referring to a descent from the periphery, we are describing almost word for word how man fell through his own weight, and how he had to leave behind him the forces and the Beings who form the clouds and the lightning—the Cherubim with flaming swords. Man falls from the Earth's periphery, out of the region where the Cherubim hold sway with their fiery swords of lightning. The ancient seers who gave us *Genesis* gazed with full powers of seership into the life of humanity weaving in the etheric heights before he fell from the regions where the Seraphim and the Cherubim hold sway.

Outer Knowledge, Inner Wisdom, Secret Teachings

The secret teachings of plasma were given to only a few because the power of plasma holds the power of life and death and cannot be given into the hands of immoral people. Dr. Rudolf Steiner could witness plasma and the etheric formative forces, which are usually invisible, through his inherent natural clairvoyance. He taught these secrets to only one person that we know of, Guenther Wachsmuth, his personal secretary. Wachsmuth wrote the book, *The Etheric Formative Forces in Cosmos, Earth, and Man* as the only exposition on ethers and plasma approved by

Rudolf Steiner. Plasma, during Steiner's day, had just been discovered (1879) and he didn't want to give out the secrets just yet, even though, in Wachsmuth's book can be found the secrets of plasma that are only now being discovered by scientism (scientific materialism). That is why we condensed Wachsmuth's book, extracted the principles, and presented the content with commentary to help point out the secrets hidden therein.

There are many primal secrets and future technologies that will come from studying plasma. Scientism is at a loss to define plasma, or for that matter, electricity, electromagnetism, nuclear energy, or the ethers that create all life on Earth and in the Cosmos. We wrote the book, *The Eternal Ethers: A Theory of Everything* to elucidate the nature, functions, and workings of the seven ethers but also felt it necessary to show that the ethers work together with plasma. Ethers and plasma are opposite poles of creation. Plasma is "not matter", and neither are the ethers. There are many similarities between ethers and plasma because plasma comes into existence using combinations of the ethers, based upon whether you are examining the Cosmos, Earth, or humanity.

Atomic-based materialism cannot allow a "non-spatial" energy or force into their incomplete cosmology based upon the fundamental forces of gravity and atoms. Scientism is not "settled science", because the theories change constantly. For instance, you might wonder why scientists do not teach people about what they think plasma is or its critical role in life. This is because, if plasma does what scientists believe they observe, then everything scientism has ever taught is null and void, useless, and meaningless data without cohesion or comprehensive application. Scientism has enslaved electricity into the marvels of the technological age and yet cannot tell anyone where electricity comes from, how it works, or explain what it is, comprehensively. Scientism is a series of changing estimations and best guesses that miss the mark when creating weapons or experimental medical treatments for humans without first thinking about the health and safety of humans. Scientism has become a new God-less religion that ignores more than they know and claims to understand things that they don't while continuously denying the wisdom and spiritual teachings of the past. Scientism works in a vacuum with one-eyed color-blindness, half-deaf perception, and no consideration for the invisible forces and beings in nature all around us. Thus, the secret teachings of plasma and the seven ethers have remained a secret to this very day as scientists continue to torture and weaponize plasma in their laboratories.

What is plasma? What will we find once the secrets of ethers and plasmas are revealed? Plasma is the Tree of Life, a name given by Steiner to chemical/sound ether and life ether working together with combinations of the warmth ether and light ether to create the entire Universe, Earth, and the human being's physical and etheric bodies.

To answer the mysterious question: What is plasma? It is almost impossible to do so without understanding the complete nature of the seven ethers. Wachsmuth's book on the ethers is the most complete presentation on the ethers that was created with Rudolf Steiner's approval. Before we endeavor to understand the ethers, we must undo and unwind some of scientism's theories that roll around in the heads of modern materialists. First, scientifically speaking, the cosmos is nothing but the fundamental forces of gravity, the strong nuclear force, and the weak nuclear force. Unfortunately, once you understand that atoms are theoretical and only exist in Earth's immediate environment which is less than .01% of the universe, then the two nuclear forces disappear.

Once you understand the secret teachings of plasma and ethers, you will realize that gravity is another "made up" force that only exists in theories. Scientism's math can't explain the 99.99% of the plasma universe with only their imaginary atoms and gravity which only approximately work in Earth's atmospheres. Science theories stop being effective when we enter the exosphere of the Earth, and certainly cannot explain the plasmasphere, the Sun/Earth magnetic reconnection, or the solar wind and solar forcing of electromagnetism, plasma, CMEs, proton winds, or any of the mechanisms that create life on Earth. Ultimately, scientism does not have a comprehensive theory of life.

Let us take a simple example of how water and oxygen come into Earth's atmospheres. Scientism theorized for a long time that all water and oxygen on Earth came from asteroids carrying them down from space where elemental (ionized) water and oxygen ions exist in abundance. Any child could figure out why that is nonsense. Eventually, scientific materialism decided on a new theory which goes something like: We don't fully understand where water or oxygen come from.

Ultimately, there is no generally accepted theory about the creation of the two most important elements necessary for life on Earth—oxygen and water. Some researchers now say that: We "believe" that water and oxygen both came from ionized elemental particles in space (the "vacuum" of space) called dust. This cosmic dust somehow organized into molecules of water and oxygen and then somehow came into Earth's atmosphere. There you have it. Scientism has only the slightest clue, but not enough wisdom and secret teachings to understand that the substances of life come from what some traditions call the "Tree of Life"; from which, flows the Waters of Life and its leaves are "for the healing of the nations" (oxygen) and its fruit gives immortality to those who can rise-up to New Jerusalem to receive these divine gifts of food and drink. Now this may sound silly to scientism, but it is an accurate analogy of what Rudolf Steiner gave as the secret teachings concerning the phenomena of life and the forces and beings that create life on Earth.

Spiritual scientific research, which we have presented above, demonstrates that ionized helium is transformed by plasma through the rays of the Sun that contain warmth, light, chemical/sound, and life ether, transforming elemental helium into ozone (O3), which protects humans from the harmful rays of the Sun. This "ozone layer" is invisible plasma that protects us and also makes oxygen from ozone in the lower atmospheres. Plasmoids in the thermosphere (6,000 degree +) take ionized helium particles of "dust" and heat them to temperatures in excess of one million degrees fahrenheit in complex dusty plasmoids to create what scientism calls "fusion" energy, the same energy they imagine drives the Sun.

The Sun's solar wind (breath) is constantly recharging the plasmasphere, which is held together by flowing along magnetic field lines called the Van Allen Belts. Plasma from the solar breath enters the plasmasphere and is recharged by a huge magnetic rope (plasma filament) every eight minutes that connects to the Sun through magnetic transfer flux portals in the Earth's magnetosphere. In this Sun/Earth symbiotic process, a magnetic reconnection portal opens that allows an Earth-wide tube of fresh plasma from the Sun to pour into the plasmasphere, expanding or contracting the parameters of the plasmasphere like human lungs breathing. This fresh breath of plasma from the Sun re-enlivens the plasmasphere with plasma that has not yet been aligned with Earth's magnetic fields, the magnetosphere, and the plasmasphere. The electromagnetic plasma flowing through these magnetic transfer flux points and portals to the Sun reconnect every eight minutes to renew magnetic re-connection that ultimately pulls the Earth behind the Sun's movement through space while feeding Earth fresh plasma. This reconnection is instantaneous, happening immediately over the 93 million miles between the Sun and Earth.

Plasma in the plasmasphere is cold and does not generally form into plasmoids until the auroral electro jet moves cold plasma from the plasmasphere into the exosphere and thermosphere thereby creating the plasma sheets of the northern lights, the aurora borealis, which meet ionized elements of carbon, nitrogen, oxygen, hydrogen, and phosphorus gases (plasma) rising up from the Earth towards the Earth's North Pole. When plasma flows into the North Pole, it aligns cold plasma with the Earth's many magnetic forces falling in through the North Pole to the Earth. This movement of cold plasma into Earth's magnetic fields drives plasma flow down into the thermosphere where complex dusty plasmoids absorb cosmic and terrestrial dust particles and spin them into electromagnetic filaments, cables, ropes, and strings that eventually create a torus shape wrapped with strands of electrical ropes. These double-layered plasmoid balls are magnetohydrodynamic fusion generators that create primal forces, both organic and inorganic.

In complex dusty plasmoid fusion cells, primal helium and hydrogen start the process of organized condensation with ozone and hydrogen peroxide, which is abundantly available in space and the Earth's exosphere, and send their "transmuted" particles into the lower atmospheres where they become primal water and oxygen. This process is continuous and was going on long before humans descended into the material/physical realm. In ancient times, this process was seen as a great tree (Yggdrasil in Norse Myths) growing into the heavens. All nine realms of gods, nature spirits, humans, ice giants, dragons, and other beings lived in different realms amongst the roots, trunk, and limbs of the World Tree. A dragon wound around the tree just as the serpent wound around the Tree of Knowledge.

In the living imagination of a World Tree that feeds and nourishes humanity, the trunk is the axis of the Earth, the roots are the south pole, and the upper branches reach up through the North Pole into the exosphere and the plasmasphere where the leaves of the tree fill the lungs of the plasmasphere with the Breath of God from the Sun. The thermosphere, filled with plasmoids, is like the fruit of the Tree of Life, and from the dew of the tree comes the Waters of Life bringing the "waters from above" to the lithosphere and the leaves bring oxygen. These "plasma secrets" are implicit in Steiner's teachings concerning the ethers.

Still, the question lingers: What is plasma? Since we have discarded atoms and the strong and weak nuclear force by showing that atoms didn't just "come into existence one day"—since the Universe is less than .01% atomic substance with an imaginary nucleus and little imaginary electrons spinning in clouds around them. Atoms, or monads, came into existence by the forces inherent in plasmas working with the ethers from the Sun to form, build, and create electromagnetically stable molecules in the thermosphere. Atoms don't really exist, they are a temporary theory. Dr. Steiner tells us that the atom is really a "monad" that replicates the electromagnetic, plasmic, and etheric mechanisms of the "life body" of the Earth as a whole. In other words, the Earth is one big primary particle, monad, or "atom" (in scientism) and every cell and monad in the human being replicates those same mechanisms. Which begs the question: Are plasmas and ethers conscious or living sentient beings, and do they live inside of me?

Plasmas and ethers are impossible to describe with terms and thoughts taken from the material world because they are non-spatial, filled with etheric geometry, seemingly universal and eternal, able to contain fusion for creating molecules, masters of levity and gravity, able to control transmutation without an "explosion", monad incubators, creators of the waters of life, invisible, and generally magical non-substance. Magic is the word humans have always used to describe what they could not understand. Plasmas and the ethers are simply magic, in a word.

Scientism tells us false fairy tales that lack imagination, creativity, or the truth of reality. Let's take the children's fairy tale called the Big Bang Theory. This unimaginative tale had to have the concept of "Inflation" added to the theory to try and explain all the gross incongruencies that facts presented to observers through their observations, experiments, telescopes, and devices. Recently, scientism had to change the "settled science" theory of the Big Bang Theory to add three new stages beyond "Inflation", called Stars III, Stars II, and Stars I. While the Big Bang was happening, when that small ball of proto-matter was exploding into the vast organized and multifarious Universe, some early stars in phase one (Stars III) came into existence that were only hydrogen and helium proto-particles that "somehow" turned into the entire amount of water we find in the Universe. Hmm. Then, stars in phase two (Stars II) came into existence (magically) and created heat-light, even though gravity and atoms had not come into existence yet. Hmm. Then, phase three happened and new stars magically came into existence (Stars I) which created matter, as we know it now, and gravity that ultimately controls the Universe. This is, of course, a different story of creation than the Big Bang Theory that was first told us, and not a single scientist has come forward to say they were wrong, or lied, or have kept lying ever since and made sure that science textbooks don't indicate past mistakes and bad theories.

Dr. Rudolf Steiner's cosmology points out that the Universe was created by the actions of spiritual beings who donated substances in the earlier "incarnations" of the Earth. We might call them Earth I, Earth II, Earth III, and now we are in Earth IV, where matter surrounds us. In Earth I, the warmth ether started moving like a nebula of invisible dust (plasma) until warmth was created in clustered balls of energy that did not glow. This is much like the plasmasphere where primal substance is not organized and swirls around as groups of large cells. Thus, Stars III should be Stars I; they got the right substances but the wrong order. Earth II worked through the forces of light ether to create primal oxygen/air. Earth III used chemical/sound ether to create a fluid environment. Earth IV used life ether to add the human being to a world of minerals, plants, and animals. Thus, warmth, light, chemical/sound, and life -ether have manifested the substances that eventually evolved into (congealed into) our earth, water, air, and fire. Thus, the four stages of the Big Bang Theory are also found in Steiner's "Incarnations of the Earth" and create the same "substances" of primal life: fire (warmth ether), air (light ether), water (chemical ether), and minerals (life ether). These ethers later evolved on Earth IV into minerals, plants, animals, and human life.

One might think that we are saying that an understanding of plasmas can only arise once an understanding of the seven ethers is obtained. This is generally true. The task of understanding ethers with science theories is doomed to fail because science does not consider that which it cannot perceive with the limited five senses, and plasmas and ethers are often invisible or imperceptible to limited human senses.

Scientism works with lab-made plasmas but has not been able to control or understand it. Spiritual science, on the other hand, prepares one to witness the four ethers that manifest on Earth, which is a good preparation for envisioning the scientific laws that plasma "breaks." We must first unwind and unlearn those old, worn-out theories of science to open our minds to cognizing plasma "substance" that is not matter, in plasma fields that are invisible, making plasmoids in etheric realms that human minds can't conceive of, and doing things materialists would call impossible. That leap into cognitive dissonance is followed by imagining the plasma wonders of quasars, black holes, millions of light-years long plasma filaments connecting galaxies, plasma clouds filled with self-organized cells that have all the characteristics of life, as well as the well-hidden plasma in the human body. Plasma and the ethers open the mind to the forces and beings of life and the primal creative spiritual beings of the Cosmos.

In many of our books, especially *The Eternal Ethers*, we reveal the magic of the four ethers, the unfallen ethers (Tree of Life), and the akasha ether (Mother of All). We show that ancient traditions have known for millennia the mysteries of ethers and plasma, both in the human "electric body" and the electromagnetic actions and creations in the Cosmos. The Earthly and Cosmic Nutrition Stream, also called the Etherization of the Blood, is explained in great detail through the workings of the etheric formative forces and also through the elements and physical mechanisms that drive the continuous feeding of the gods and their reciprocal feeding of humans. Suffice it to say, the same processes of the Earth are duplicated in a smaller version in the human being. For example, the Etherization of the Blood is very similar to plasma flowing into the magnetic north pole creating the aurora borealis. From heart to head, ether streams carry invisible plasma (electromagnetic energy as the combustion of ionized elemental substance) up to the pineal gland where an internal aurora is created around the corpora quadrigemini and pineal gland stimulating piezoelectric discharges from the calcium carbonate crystals in the pineal body.

When discharges from the pineal gland enter the cerebral spinal fluid in the fourth ventricle, which is surrounded by the strongest electromagnetic field in the brain located in the cerebellum, the discharge moves through the third ventricle's fluid that is filled with minerals and metal particles descending from the first and second ventricles into the cerebral spinal fluid filling the third ventricle. This piezoelectric discharge causes the third ventricle to "light up" with a piezoelectric charge that stimulates the pituitary gland causing it to secrete pituitrin into the blood and nerve systems of the body. This process is a plasmatic magnetohydrodynamic generator that creates the bio-electromagnetic field emitted by the head.

We can see in the comparison of the human head and the Earth's North Pole that the action of plasma and the ethers working together create a display of light and color in the aurora and that a white "halo" around the North Pole is similar to the plasma rising in the etheric stream from heart to head that creates a rainbow around the pineal gland and an oval glowing light (halo) around the head of a spiritual person that a clairvoyant can see. This is described in great detail by many spiritual traditions that have secret teachings concerning these forces and the beings who create them. We can also see in the Tree of Life analogy the plasmodic creation of water and oxygen which are the very processes of spiritual nutrition that can be found in ancient traditions concerning spiritual nourishment. The clearest traditions are those that concern the human heart.

Cosmically speaking, plasma is gravity, and gravity is life ether which can act in a cosmic or earthly manner sucking matter to its being, holding "matter" together. Gravity creates the pulse beat of the contracting heart which drives the motion of the heart as the heart "falls" into the suction force of Earth's supposed gravity, or life ether. The human heart, which is the biggest dynamo of bio-electromagnetism in the human body, is another magnetohydrodynamic generator that produces a large magnetic field that extends out about 10-12 feet from the human body. The plasma generator in the head, that we described above, is at least 50 times weaker than the heart.

The third magnetohydrodynamic (MHD) generator in the human body has now been found in the Enteric Nervous System (ENS), in the lining of the intestines where the longest peripheral nerve in the body is found. The ENS is called the "second brain" which can create coherence with the heart and brain generators via the vagus nerve, the longest nerve in the body (10th cranial nerve). These three generators must be synchronized and coherent for the human being to bring into balance their capacities of thinking, feeling, and willing. All three generators are plasma generators that create plasmodic bio-electromagnetism that feeds the nerves, blood, tissue, and all aspects of the human body that combine to create a platform for individualized self-awareness and the ability to consciously reflect upon one's thinking, feeling, and willing. All three generators can be explained by scientific evidence, facts, data, and measurements, but they still cannot be seen, explained, or understood by scientism. Many spiritual paths describe these mechanisms in extensive detail and explain what these forces are, and WHO they are. What they all agree upon is that these cosmic "forces" are not the "Beings" behind them. Some type of invisible, cosmic Being donates and controls these forces with "intelligent design" that looks a great deal like "life" itself. In other words, spiritual beings work through the forces of ethers and plasmas, but they are not those forces—they are emanating those forces, like a person in the three-dimensional world can cast a shadow into a two-dimensional world. There are big differences between Beings and forces.

If ethers and plasmas are the forces that came from Beings, then who are the Beings behind these forces that mold and shape matter on Earth and in the cosmos? Obviously, the forces are invisible, but we can work with and manipulate those forces, then the Beings themselves are invisible and almost unimaginable to childlike humanity. If we are to believe in scientism, these forces have been around for about four or five billion years slowly building the world we see with our limited senses. Humans are in a nursery being incubated like the cosmic particles of helium and hydrogen in space are birthed into inorganic life in the atomospheres around the Earth. Beings work behind and through plasmas and ethers as they descend into matter, like water crystalizing by the force of cold. Matter is precipitated over four stages of condensation into the kingdoms of nature: mineral, plant, animal, and human. Humans are like the fruit of the Tree of Life dropping from the Garden of Eden into the world of cold, hard matter. Once the fruit of the Tree of Knowledge of Good and Evil was picked, the garden disappeared along with the Tree of Life that elevated itself to a place where humans cannot pick its fruit so easily.

The Tree of Life is still in the Garden of Eden on the White Island that the aurora borealis creates above the North Pole. A conscious spiritual aspirant can cross the "rainbow bridge" into that realm once the seeker understands the other eight realms in the World Tree and the processes of the Etherization of the Blood, or the Earthly and Cosmic Nutrition Stream.

The realm of the gods at the top of the World Tree Yggdrasil was called Asgard, where the Norse gods and goddesses lived, far above the land of humans called Midgard. Asgard can be imagined as the realm of unfallen chemical/sound and life ether that all humans pass through when they go to sleep and pass into the outer realms of the planets.

Most people are unconscious during sleep when they pass through these realms. Initiates are awake in their sleep and understand that they go into these etheric/plasmic realms at night like an inquisitive student going to school or a hungry person going to eat. The rotating white aurora (the White Island) is the portal to spiritual realms that are far beyond what we have described so far. This is the realm called Shambhala, Mount Meru, Tushita Heaven, New Jerusalem, the White Island, the Temple of Wisdom, the Hall of Wisdom, and many other names. It is the realm of plasma/ether which is timeless, non-spatial, filled with consciousness and archetypes, and defies all Earth-bound knowledge based on the material world. It is a realm called the astral light which is not bound by matter.

After all of these unimaginable ideas have been crudely presented with Earth-bound words, we need to now consider the genesis and primal aspects of the ethers and plasmas. The primal ether was the Mother of all ethers called akasa, akasha, Aviditi, Vac, Fohat, solar magnetism, the Mother Lodge of the Sun, the Solar Breath, and many other names. It is the unmanifest that is manifest, the Alpha and Omega, the original ground of being, spaceless space and timeless time with a consciousness born before consciousness. Many call akasha ether the Akashic Records, where everything that ever happened is recorded and can be accessed by moral initiates.

Akasha was the first mover, the original breath, the magic word AUM, the Word, the Logos, and the original creator. Akasha, according to ancient teachings, gave birth to warmth, light, sound, and life in their original forms. Akasha is all around us, but it is nowhere. There is no explanation that can describe akasha; it is formless form and the finite vessel that holds the infinite. It is the Holy Grail of primal space that suffers creation and is the force that started the original spiral movement in the ocean of primal material.

Just as mystical and sacred to the ancients as plasma, were the unfallen ethers of chemical/sound and life ether. These two ethers from the Sun were seen as life itself. Usually, the ancients only saw the power of the Sun in chemical/sound and life -ether when there was a cataclysm with plasma on the Sun. They understood these two forces to be the greatest powers in the Cosmos. Fortunately, solar coronal mass ejections and solar cataclysms happen infrequently, but they were not forgotten as the force of the chemical/sound and life -ether bombarded the Earth. The ancients respected the Sun and the Beings who live there and knew that a Solar God watches over the Earth to make sure the power of the Sun would only gradually be absorbed into the human being. This loving Solar God was found in the teachings of every culture and religion. Humans inherently knew that the Sun, who created us, would not abandon us. Therefore, to connect to the life-giving ethers and plasmas of the Sun, the solar deity called Christ descended to Earth and connected the Sun to the Earth for all time to come; not unlike the breathing Sun/Earth magnetic reconnection that happens every eight minutes and has birthed life on Earth.

It was this teaching concerning the Solar System that was secret in ancient times. Before Jesus of Nazareth became the vehicle for this Solar Deity, all religions had teachings about the descent of a god from the Sun, in one form or another. Probably the best pre-Christian personification of the Love of the Sun was the ancient Indian belief in Fohat, a solar magnetic force of fire that reaches into the heart of humans and sets the spark of spirit on fire. This being, called Fohat, was around since the creation of the Cosmos and the idea of the human being was planned and created in the Mother Lodge of the Sun by Christ, the Second Person in the Holy Trinity, and the guiding beings of the Second Rank of spiritual hierarchical beings called: Spirits of Wisdom, Spirits of Motions, and Spirits of Form. Fohat was a personal god who descended on the rays of the Sun and found a home in the human heart. Fohat is inexplicable, inexpressible, unnamable, cannot be contained, and moves through all worlds like a blazing fire.

Thus, we can see that no book can contain the full description of the types of plasma and how they function. Even a comprehensive understanding of the ethers, both fallen and unfallen, only prepares you to understand plasmic electromagnetism and the mechanisms by which it helps the etheric formative forces bring life to birth and sustain it with earthly and cosmic nutrition. We hope that our efforts here will spark a flame of unquenchable desire to study diligently to understand the forces and beings who have worked so hard to provide realms where individualized human consciousness can come to birth and find freedom and love everywhere. As Rudolf Steiner said: The Cosmos is filled completely with an unimaginable amount of love.

The Battle Between Light and Dark

We are engaged in a mighty battle of Light versus Darkness fighting for the soul and spirit of humanity, regardless of whether we are conscious or not that the "War in Heaven" has "fallen" to Earth. The luciferic hosts from the celestial hierarchies fell to Earth as a regression in their own spiritual development. This fall was due to the selfishness of the luciferic hosts, which was birthed by their jealousy of and lust for, the power of God. Selfishness cannot exist in the spiritual world, and thus, Lucifer's desires created the necessity for the prideful fall of one of the highest angels in heaven. Lucifer's light is selfish, self-effulgent, non-radiant, and devoid of the magical substances and power of God's true light that is found in the Sun's warmth, light, harmony, and life.

Lucifer's light lacks the spiritual nourishment that humans derive from solar light, which is ruled by the Second Logos of the Holy Trinity—Christ. The Archangel Michael cast Lucifer out of heaven. Christ and Michael battle Lucifer's deluded minions on Earth that try to lure human thinking into the same luciferic, selfish, downward path into "dark light." This evil darkness is alive and is the antithesis of Christ's light that gives life. Lucifer's light leads into caves of delusion and mental illness—which is one of the earmarks of materialistic scientists who are following the same path of perdition that Lucifer and his hosts followed.

Ancient civilizations described the battle between living light and living darkness in many ways. The ancient Greeks believed that after visiting the healing temple of Asclepius, sleeping overnight, and reporting their dreams to the temple priest, the new dawn brought the "light of Helios" healing their illness that was seeded by the living darkness of night. After the dreams revealed the nature of the illness, it was sunlight (Helios) that healed the illness. Illness was bred in darkness and healed in the light.

In ancient Persia, the priest-king Zarathustra taught that the light of the Sun was a being named Ahura Mazdao, who lived on the Sun but, at that time, could be encountered in the living rays of the Sun shining down into the darkness of the Earth. The darkness of the Earth was ruled by the Being of Darkness called Angri Manu, or Ahriman. Ahriman was seen as the collective forces of darkness that ruled illness, old age, entropy, and death. The battle between these dual forces of Light and Darkness was continuous, and the prize was the soul and spirit of each human being.

In modern times, misguided materialistic scientists only believe temporarily in their limited theories and have an aversion to attributing "life" to light, or basically anything else. These unimaginative and immoral theories lack wisdom and are, in fact, inspired by living Beings of Darkness who have "fallen" from the ranks of the spiritual hierarchies. The "unfallen" spiritual hierarchies selflessly donate their substance to create the Earth, and all natural materials we see in the Cosmos, by slowing down the frequency of light enough to crystalize into material substance. Contrary to the self-sacrifice of the good hierarchical beings, fallen beings of darkness, luciferic rebels, selfishly try to steal the spiritual ego of human beings to build a world of evil in sub-natural realms by capturing light and denaturing its living substance into its opposite. Both of these battling worlds, light and dark, are generally invisible to human sense perception, which science limits to the five materialistic senses, ignoring the other seven senses.

Humans are beings of love and freedom who must make a choice between Living Light—filled with wisdom, or Cold Darkness that lures the soul into selfish pursuits of delusion, soul-illness, and spiritual death. It is a challenge of our apocalyptic times to decide to either become more conscious and evolve into the planned spiritual ascension of the soul and spirit, or be mesmerized by the dark magic of mad-scientists and demonic doctors who try to selfishly play god with future human evolution.

It seems wonderful, fresh, and delightful that scientists now claim that they can perform god-like medical procedures with "light" that can alter DNA, RNA, and the natural processes of the human body. They call it "photonic medicine", and doctors now have numerous medical treatments available that address issues ranging from skin and teeth problems to biomodulation of brain neural pathways and DNA manipulation. To date, none of these treatments have been approved, and there has not been adequate research to determine the side-effects. Under the auspices of the goodness of "light", which is generally seen as a healthy thing, mad-scientists have immorally snuck "living darkness" evil intentions into these supposedly harmless treatments that utilize man-made light. In fact, there is no end to the problems that "wannabe-god" doctors and mad-scientists have already created with photonic medicine.

In an effort to shed spiritual light on this impending disaster of pseudo-medicine, we present below some of the non-approved medical treatments that immoral doctors, using untested light-treatments, are now foisting on humans without informed consent or any concern for the health and wellbeing of their patients. Of course, this is nothing new to allopathic medicine that is responsible for the third highest cause of death in America—iatrogenic death (death by doctor). Many would argue that doctor induced death is actually the number one cause of death in America. Reading about the mad-scientists' medical treatments (experimentations) involving light, which are enumerated below, will give a sample of where this new Frankenstein Medicine is headed. It is the same place many other types of treatments and drug therapies have already gone—into the hands of evil beings who wish to destroy humanity, not heal it.

Photomedicine

Photomedicine is a type of biotechnology that specializes in the therapeutic application of light, often referred to as "light medicine." It uses non-ionizing electromagnetic radiation in dermatology, oncology, surgery, radiology, DNA manipulation, and diagnostics. The fundamental particle of light is referred to as a photon. Photons are both a particle and a wave. They are deemed to carry no charge and have no mass. In a biological system, photons are emitted when electrons move from one energy state to another. Photomedicine is used to treat various conditions, such as tissue injuries, psoriasis, seasonal affective disorder, circadian rhythm disorders, cancer, tumors, and alopecia.

Weak emission of light from cells in a living organism were discovered by the Russian embryologist Alexander Gurwitsh in 1926, who called them "mitogenetic rays." Half a century later, the German researcher Fritz Albert Popp, a

Nobel Prize nominee in Physics, reconfirmed their existence and established the term "biophoton." Popp experimentally demonstrated that dozens of photons of light are emitted every second from every square centimeter of biological substance. Popp demonstrated that biophoton emission is not limited to thermal radiation or bioluminescence.

Light Therapy

Light therapy, also known as phototherapy or classically referred to as heliotherapy, is a method recognized by scientific medicine for treating various diseases. It includes exposure to outdoor daylight or specific indoor artificial light sources.

The care guideline for unipolar depression recommends light therapy, especially for depression that follows a seasonal pattern (seasonal affective disorder). There is tentative evidence supporting its use for treating depressive disorders that are not seasonally dependent. As a treatment for disorders of the skin, the second kind of light therapy, called ultraviolet light therapy, is meant to treat neurodermatitis, psoriasis, acne vulgaris, eczema, and neonatal jaundice.

Photodynamic therapy is a form of phototherapy that uses nontoxic light-sensitive compounds selectively exposed to light, making them toxic to targeted malignant tumors and other diseased cells.

Some light therapy involves using a light box that emits up to 10,000 lux of light at a specified distance, which is much brighter than a customary lamp, or a lower intensity of specific wavelengths of light from the blue (460 nm) to green (525 nm) areas of the visible spectrum. A 1995 study showed that green light therapy at doses of 350 lux produces melatonin suppression and phase shifts equivalent to 10,000 lux white light therapy, but another study suggests that the blue light often used for SAD (seasonal affective disorder) treatment should perhaps be replaced by green or white illumination, because of a possible involvement of the cones in melatonin suppression.

Contraindications to light therapy for Seasonal Affective Disorder include conditions that might make the eyes more vulnerable to phototoxicity, tendency toward mania, photosensitive skin conditions, or use of a photosensitizing herb (such as St. John's wort) or medications. Patients with porphyria should avoid most forms of light therapy. Patients taking certain drugs, such as methotrexate or chloroquine, should use caution with light therapy, as these drugs could potentially cause porphyria.

Side effects of light therapy for Sleep Phase Disorders include jumpiness or jitteriness, headache, eye irritation, and nausea. Some non-depressive physical complaints, such as poor vision and skin rash or irritation, may improve with light therapy.

Light therapy treats the following concerns: vitamin D deficiency, skin conditions, acne vulgaris, atopic dermatitis, cancer, psoriasis, vitiligo, wound healing, retinal conditions, seasonal affective disorder, non-seasonal depression, chronic Circadian Rhythm Sleep Disorders, jet lag, situational CRSD, Parkinson's and Alzheimer's sleep disorders, and neonatal jaundice.

Diagnostic Photomedicine

Diagnostic photomedicine involves the use of light for various imaging technologies, such as X-ray, MRI, and PET. Other forms of photomedicine include photodynamic therapy (PDT), low-level laser therapy (LLLT), and LED therapy. Photodynamic therapy (PDT) is a treatment that uses light-sensitive medicine and a light source to destroy abnormal cells. It can be used to treat some skin and eye conditions, as well as certain types of cancer.

Photodynamic therapy can treat abnormal cells in parts of the body accessible to a light source, such as the skin, eyes, mouth, food pipe, esophagus, and lungs. Common conditions treated with PDT include actinic keratoses, Bowen's disease, basal cell carcinoma, macular degeneration, esophageal cancer, mouth cancer, lung cancer, warts, acne, and Paget's disease.

When some photomedicines, such as templated graphene oxide, are exposed to light, they activate and cause a reaction that damages nearby cells. This allows small abnormal areas of tissue to be treated without the need for surgery. Infrared light can heal cells and tissue, whereas ultraviolet light kills cells and tissue and is particularly effective at "turning off" targeted genes using luciferase and luciferin. Bioluminescence results from the oxidation of luciferin and the enzyme luciferase. Using light to turn DNA on and off impacts the fields of DNA biotechnology, synthetic biology, and epigenetics.

Controlling the Brain Through Light

After several independent studies showed that living cells not only just radiate light but also absorb light, scientists are now investigating a new form of communication using light. Biophotons could provide an additional signaling pathway alongside electrical and chemical pathways for intra- and intercellular communication. It is known that photosensitive biomolecules in cells and neurons can absorb biophotons and transfer the absorbed biophoton's energy to nearby biomolecules through resonant energy transfer, which can induce conformation changes and trigger complex signal processes in cells and between cells. Indeed, hollow microtubules with constant inner diameters inside the brain could act as optical fibers for biophoton transmission within brain nerve cells.

It has been suggested that the major source of biophotons is DNA. The first supporting fact is that cells emit biophotons even when the cytoplasm is damaged, however, when the nuclei is removed, biophoton emission stops. Red blood cells, which have no active chromatine (DNA) are the only cells that do not emit biophotons. The mechanism of biophoton absorption, storage, and emission is, however, not well understood.

Photomedicine has been particularly effective at regenerating frayed DNA telomeres. This ability to repair DNA suggests potential future medicines. Some of the most effective pain relief comes from the application of LED light, infrared light, and laser light directly to the affected area. These devices are over the counter "light medicine." In effect, light heals, and the use of light-absorbing graphene oxide, in all of its forms, kills.

Photobiomodulation

Low-level laser therapy (LLLT), cold laser therapy, or photobiomodulation (PBM) is a form of medicine that applies low-level (low-power) lasers or light-emitting diodes (LEDs) to the surface of the body. Whereas high-power lasers are used in laser medicine to cut or destroy tissue, it is claimed that the application of low-power lasers relieves pain or stimulates and enhances cell function. The effects appear to be limited to a specified set of wavelengths, and administering LLLT below the dose range does not appear to be effective. Photochemical reactions are well known in biological research, and LLLT utilizes the first law in photochemistry: light must be absorbed by a chemical substance in order for a photochemical reaction to take place. In LLLT, that chemical substance is represented by the respiratory enzyme cytochrome c oxidase, which is involved in the electron transport chain in mitochondria.

Various LLLT devices have been promoted for treating several musculoskeletal conditions, including carpal tunnel syndrome (CTS), fibromyalgia, osteoarthritis, and rheumatoid arthritis. They have also been promoted for temporomandibular joint disorders, wound healing, smoking cessation, and tuberculosis. LLLT seems effective in preventing oral mucositis in recipients of a stem cell transplant with chemotherapy. In other areas, evidence for LLLT remains conflicted. Some studies suggest that LLLT may be modestly effective relieving short-term pain for rheumatoid arthritis, osteoarthritis, hair loss, brain injuries, cancer treatment, chronic low back pain, acute and chronic neck pain, tendinopathy, and possibly chronic joint disorders. The evidence for LLLT being useful in dentistry and in the treatment of wound healing is experimental.

Brain Photobiomodulation

Brain photobiomodulation (PBM) therapy using red to near-infrared (NIR) light is an innovative treatment for a wide range of neurological and psychological conditions. Red/NIR light is able to stimulate Complex IV of the mitochondrial respiratory chain (cytochrome c oxidase) and increase ATP synthesis. Moreover, light absorption by ion channels results in the release of Ca2+ and leads to the activation of transcription factors and gene expression. Brain PBM therapy enhances the metabolic capacity of neurons and stimulates anti-inflammatory, anti-apoptotic, and antioxidant responses, as well as neurogenesis and synaptogenesis. Its therapeutic role in disorders such as dementia and Parkinson's disease, as well as to treat stroke, brain trauma, and depression, has gained increasing interest. In the transcranial PBM approach, delivering a sufficient dose to achieve optimal stimulation is challenging due to the exponential attenuation of light penetration in tissue.

Alternative approaches, such as intracranial and intranasal light delivery methods, have been suggested to overcome this limitation.

Fluorescence Biomodulation

Fluorescence biomodulation is a form of photobiomodulation, which utilizes fluorescence energy to induce multiple transduction pathways that can modulate biological processes through the activation of photo-acceptors found within many different cell and tissue types. Photo-acceptors are molecules that do not explicitly specialize in light absorption but possess the ability to do so in the presence of light which in turn can enhance their functioning.

To generate fluorescence, specialized light absorbing molecules (chromophores) are employed to translate light energy into a high-energy emissions of fluorescence through a mechanism known as Stokes shift. Fluorescence biomodulation differs from photobiomodulation in that it employs fluorescence as a photo vehicle to induce biomodulation. Fluorescence, as generated by chromophores, is displayed as a broad spectral distribution of wavelengths and/or frequencies which can be controlled to penetrate tissues to various degrees. Tailoring fluorescence biomodulation allows compatibility between the specific emissions of fluorescence and the unique light absorbing characteristics of different cell and tissue types in the body. Shorter wavelengths (<600 nm) within the visible spectrum cannot penetrate deep into tissue and are localized within the epidermis or dermis. Conversely, longer wavelengths (>600 nm) within the visible spectrum penetrate further up into the hypodermis.

Blood irradiation therapy is an alternative medical procedure in which the blood is exposed to low level light (often laser light) for therapeutic reasons. The practice was originally developed in the United States, but most recent research on it has been conducted in Germany (using UV lamps) and in Russia (in all variants). Low-level laser therapy has been tested for a wide range of conditions, but rigorous double-blind studies have not yet been performed. Furthermore, it has been claimed that ultraviolet irradiation of blood kills bacteria by DNA damage and also by activation of the immune system.

Red Light Therapy

Red light can reverse a condition called mitochondrial dysfunction, which is the root cause of rapid aging. Mitochondrial dysfunction occurs when the mitochondria, the energy producers of the cells, are unable to convert raw materials into energy. This condition has been flagged as a leading contributor to many chronic conditions, including neurodegenerative disorders and rapid skin aging since energy-depleted cells are unable to perform their functions and become more vulnerable to viruses and bacteria.

Deep penetrating red light therapy restores normal mitochondrial functioning by stimulating the production of a cellular fuel called adenosine triphosphate (ATP). More fuel energizes the cells so they can perform their specialized functions, repair themselves, and replicate.

Cellular senescence (the cell's inability to progress through the cell cycle) is a normal process that happens as we age. The reason that cellular senescence occurs is due to our telomeres shortening over time. Each time cells divide for growth or tissue repair, telomere length reduces. When it reaches zero, the cells can no longer divide and die. Red light therapy, in conjunction with maintaining a healthy lifestyle, will delay telomere shortening and potentially help them regrow.

The negative effects of red-light therapy include mild to moderate pain, peeling skin, skin redness, crusting of the skin, tightness, swelling, itching, and acne flare-up.

Dangers

Ultraviolet light causes progressive damage to human skin and erythema even from small doses. This is mediated by genetic damage, collagen damage, as well as destruction of vitamin A and vitamin C in the skin and free radical generation. Ultraviolet light is also known to be a factor in the formation of cataracts. Ultraviolet radiation exposure is strongly linked to the incidence of skin cancer.

Optical radiation of any kind with enough intensity can cause damage to the eyes and skin including photo-conjunctivitis and photokeratitis. Researchers have questioned whether limiting blue light exposure could reduce the risk of age-related macular degeneration.

Light exposure is reported to suppress the production of melatonin, which affects our body's circadian rhythm and can decrease sleep quality. It is reported that bright light therapy may activate the production of reproductive hormones, such as testosterone, luteinizing hormone, follicle-stimulating hormone, and estradiol.

Light therapy is a mood-altering treatment, and just as with drug treatments, there is a possibility of triggering a manic state from a depressive state, causing anxiety and other side effects. While these side effects are usually controllable, it is recommended that patients undertake light therapy under the supervision of an experienced clinician, rather than attempting to self-medicate.

Contraindications to light therapy for seasonal affective disorder include conditions that might render the eyes more vulnerable to phototoxicity, tendency toward mania, photosensitive skin conditions, or use of a photosensitizing herb or medication. Patients with porphyria should avoid most forms of light therapy. Patients on certain drugs such as methotrexate or chloroquine should use caution with light therapy as there is a chance that these drugs could cause porphyria. Side effects of light therapy for sleep phase disorders include jumpiness or jitteriness, headache, eye irritation and nausea.

Because photobiomodulation was shown to stimulate the growth of cancer cells in cell culture studies and can also increase the aggressiveness of some cancer cells, some commentators have asserted that PBM may be contraindicated in clinical use in patients with cancer.

Blue light therapy can potentially damage skin in a number of ways:

- Induces oxidative stress, which can lead to inflammation and collagen degeneration
- Disrupts activity of melanocytes, which leads uneven and excessive pigmentation
- Contributes to photoaging, or skin aging induced by light waves (similar to sun exposure)
- Weakens the outermost layer of your skin and delays its recovery from daily skin stressors
- Reactive oxygen species damage DNA and cause breakdown of collagen and elastin, resulting in skin laxity, wrinkles and accelerated aging.
- Interfere with your circadian rhythm and disturb sleep. It's well-known that sleep deprivation can cause changes to the skin, including dullness, swollenness, redness, and the ever-dreaded dark under-eye circles.

The Mysteries of Meteorites and Comets

Rudolf Steiner's relationship with meteorites can be summed up in the phrase: Get meteorites and work with them. Whether the spiritual scientist wishes to follow this injunction and aid the Archangel Michael, the spirit of the Sun, in his battle against the Dragon, Ahriman, or focus on creating medicaments of meteoric iron to enhance the processes of combustion in the human blood, the power of meteorites should be recognized as unique, powerful, and beneficial to humanity. In Anthroposophic Medicine, meteoric iron is used to strengthen the ability for red blood cells to take up oxygen and enhance the process of respiration through a "slow burning fire." Meteoric iron also can be used for meditation and contemplation on the cosmic workings of our solar system. Meteors and comets inspire the aspirant to consider their source and nature as well as the meaning of their place in the balance of our solar system. They can also cause undue fear and consternation in naive individuals as they fall to Earth with a potential power comparable to an atomic explosion. Balancing wonder and awe while watching a celestial meteor speed towards Earth with its fiery tail streaming behind it, against the fear and terror of a potential cataclysm is reminiscent of the image of the Archangel Michael casting Lucifer from Heaven to Earth. A single large meteorite striking Earth could cause a disaster that wipes out most life in an instant through a catastrophic explosion. The Archangel Michael subduing the Dragon is not only an analogy but also a threatening reality. This cosmic imagination is active both in our solar system and in the bloodstream of the human body.

Thousands of Near Earth Objects, such as comets and meteors, approach Earth at varying frequencies, seemingly appearing out of nowhere and passing by with the potential to impact humanity's fate. Often, asteroids pass so close to the Earth that they enter the realm between the Moon and Earth. Some meteors seem to come out of nowhere, and scientists admit they "did not see them coming." These cosmic wanderers are often obscured by the Sun, and their origin is still unknown to the scientists whose job it is to monitor and track these possible cosmic disasters. Rudolf Steiner indicated that many iron meteorites originate directly from the Sun, expelled from its core through sunspots. This insight from Steiner is only now being understood after many years of scientific misunderstanding regarding the Sun and its relationship with the Earth. The spirit of the Sun,—Christ, the Elohim, and the Archangel Michael—is still casting the Old Dragon to Earth, as the *Bible* described. These harbingers of possible disaster hold the fate of humanity in their unpredictable hands, as it were. Humanity lives in a symbiotic relationship with the Sun and its messengers—shooting stars, meteorites, and comets—as human consciousness interacts with the dynamic and mysterious functions of the Sun. This may be difficult to understand for those who forget that the Sun is the Mother of all life on Earth and certainly She does not wish to harm Her children in any undue manner. But it is incumbent upon Her children to study Her nature and actions with wonder, awe, and respect if they wish to understand Her mysterious ways.

Rudolf Steiner has told us in his teachings that all things in the Cosmos are also found in the human physical constitution, except for comets. However, comets do have a physical analogy to the forces of pregnancy in expectant mothers. Generally, comets originate from the Oort Cloud at the edge of our solar system, emerging into physical manifestation from a realm poorly understood by modern science. Comets seem to appear from "nothingness" and follow a periodic path around the Sun and then return to the Oort Cloud and disappear from physical manifestation. As such, they seem to "incarnate" and "excarnate" in an inexplicable manner that scientists can only speculate about. Additionally, the fact that an "icy" comet does not melt despite often coming very close to the Sun during its perihelion adds another layer of mystery. A comet's tail always streams away from the Sun, regardless of whether the comet is approaching or receding. This phenomenon puzzled scientists until it led to the discovery of solar winds, the Sun's breath, which was only "discovered" in modern times. Thus, comets are another messenger of the Sun that wishes to educate human beings and clean the solar system of poisons.

Rudolf Steiner has explained many of these cosmic mysteries in his course on astronomy, which was given to the first Waldorf school teachers, along with many other curious and misunderstood concepts concerning comets, the Sun, solar light, and meteors, among many other topics. He also indicated that cyanide is prevalent in the tails of comets, which only recently has been discovered, confirming again his advanced understanding of our solar system and its workings. According to Steiner, comets appear periodically as a mechanism of cleaning our solar system of poisons dangerous to humanity. His insights concerning comets should teach us that many mysteries that Spiritual Science (Anthroposophy) has revealed will only be fully understood by scientists in the future as new discoveries come to light that prove Steiner's revelations to be true. Thus, whether we study comets, meteors, meteorites, our Sun, or planetary influences, we should lend a listening ear to what Anthroposophy has to say about the true realities of our Cosmos that are yet to be understood.

Meteorites

A meteorite is a rock that falls to Earth from space, often from the asteroid belt, which is the remnant of a planet that has broken into many pieces in the ancient past. Most are far older than the Earth and they provide some of the only samples we have of other worlds in our solar system, whether planets, asteroids, or comets from the Oort Cloud. Some meteorites even contain tiny particles that formed in other stars that existed long before our Sun. The study of meteorites has helped us understand the beginnings of our solar system, how planets and asteroids formed, and how impacts of large meteorites have altered Earth's history and life on our planet.

Meteorites come from inside our solar system. Most of them are fragments of asteroids that broke apart long ago in the asteroid belt, located between Mars and Jupiter. Such fragments may orbit the Sun for eons before falling to Earth, while others come directly from the Sun and are cast out of sunspots, often directly towards the Earth.

Meteorites can weigh up to around sixty tons, or they can be as small as a grain of sand. The vast majority of meteorites are fragments of shattered asteroids, which are often pulled out of the asteroid belt by the force of Jupiter. A small number of meteorites are pieces of rock from the surfaces of other planets like Mars, Mercury, and the Moon. Certain primitive meteorites contain the first solid material to form in our solar system and provide clues to the nature of the elements present in the solar system as a whole.

There have been nearly 1,100 recovered falls (meteorites seen to fall) and nearly 40,000 finds (found but not seen to fall). It is estimated that probably 500 meteorites reach the surface of the Earth each year, but less than 10 are recovered. Most meteorites fall into the ocean or on land in remote areas of the Earth. Many small meteoroids pass close to Earth each day. It is estimated that each day one or two 5–10-meter objects pass within the Moon's orbit and that there are probably 30 million Near-Earth Objects that could become a "falling star" that reaches the Earth.

> *"The first angel sounded, and there followed hail and fire mingled with blood, and they were cast upon the earth: and the third part of trees was burnt up, and all green grass was burnt up. And the second angel sounded, and as it were a great mountain burning with fire was cast into the sea: and the third part of the sea became blood; and the third part of the creatures which were in the sea, and had life, died; and the third part of the ships were destroyed. And the third angel sounded, and there fell a great Star from Heaven, burning as it were a lamp, and it fell upon the third part of the rivers, and upon the fountains of waters; and the name of the star is called Wormwood: and the third part of the waters became wormwood; and many men died of the waters, because they were made bitter."*
>
> *Revelation* (Chapter 8:7-11)

The Michael Imagination

***Four Seasons and the Archangels*, Rudolf Steiner, Lecture I, October 5, 1923, GA 229**

"This sulfurizing process in human bodies at midsummer, although it is so mild and gentle and imperceptible to man himself, has very great importance for the evolution of the cosmos. A great deal happens out there in the cosmos when, in summer, human beings shine inwardly with the sulfur process. Seen from other planets, the inner being of man then begins to shine, becoming visible as a being of light to the etheric eyes of other planetary beings."

"From the standpoint of the cosmos, this is a majestically beautiful sight, for it is in glorious astral light that human beings shine out into the cosmos during high summer, but at the same time, it gives occasion for the Ahrimanic power to draw near to man. For this power is very closely related to the sulfurizing process in the human organism. We can see how, on the one hand, man shines out into the cosmos in the St. John's light, and on the other, how the dragon-like serpent-form of Ahriman winds its way among the human beings shining in the astral light and tries to ensnare and embrace them, to draw them down into the realm of half-conscious sleep and dreams. Then, caught in this web of illusion, they would become world-dreamers, and in this condition, they would be prey to the Ahrimanic powers. All this has significance for the cosmos also."

"When, in high summer, from a particular constellation, meteors fall in great showers of cosmic iron, then this cosmic iron, which carries an enormously powerful healing force, is the weapon which the gods bring to bear against Ahriman, as dragon-like he tries to coil round the shining forms of men. The force which falls on the earth in the meteoric iron is indeed a cosmic force whereby the higher gods endeavor to gain a victory over the Ahrimanic powers when autumn comes on. This majestic display in cosmic space, when the August meteor showers stream down into the human shining in the astral light, has its counterpart—so gentle and apparently so small—in a change that occurs in the human blood. This human blood, which is in truth not such a material thing as present-day science imagines but is permeated throughout by impulses from soul and spirit, is rayed through by the force which is carried as iron into the blood and wages war there on anxiety, fear, and hate. The processes which are set going in every blood-corpuscle when the force of iron shoots into it are the same, on a minute human scale, as those which take place when meteors fall in a shining stream through the air. This permeation of human blood by the anxiety-dispelling force of iron is a meteoric activity. The effect of the raying in of the iron is to drive fear and anxiety out of the blood."

"As the gods with their meteors wage war on the spirit, who would like to radiate fear over all the earth through his coiling serpent-form, and while they cause iron to stream radiantly into this fear-tainted atmosphere, which reaches its peak when autumn approaches or when summer wanes—so the same process occurs inwardly in man when his blood is permeated with iron. We can understand these things only if we understand their inner spiritual significance on the one hand, and if, on the other, we recognize how the sulfur-process and the iron-process in man are connected with corresponding events in the cosmos."

"A person who looks out into space and sees a shooting-star should say to themself, with reverence for the gods: "That occurrence in the great expanse of space has its minute counterpart continuously in myself. There are the shooting-stars, while in every one of my blood-corpuscles iron is taking form: my life is full of shooting-stars, miniature shooting-stars." And this inner fall of shooting-stars, pointing to the life of the blood, is especially important when autumn approaches, when the sulfur-process is at its peak. For when humans are shining like glow-worms in the way I have described, then the counterforce is present also, for millions of tiny meteors are scintillating inwardly in their blood."

"This is the connection between the inner man and the universe. And then we can see how, especially when autumn is approaching, there is a great raying-out of sulfur from the nerve-system towards the brain. The whole person can then be seen as a sulfur-illuminated phantom, so to speak."

"But raying into this bluish-yellow sulfur atmosphere come the meteor swarms from the blood. That is the other phantom. While the sulfur-phantom rises in clouds from the lower part of man towards his head, the iron-forming process rays out from his head and pours itself like a stream of meteors into the life of the blood."

"Such is humanity when Michaelmas draws near. And one must learn to make conscious use of the meteoric force in one's blood. One must learn to keep the Michael Festival by making it a festival for the conquest of anxiety and fear; a festival of inner strength and initiative; a festival for the commemoration of selfless self-consciousness."

"So, at Michaelmas—if the Michael Festival is to be rightly understood—we must celebrate that which lives spiritually in the sulfurizing and meteorizing process in humanity and should stand before human consciousness in its whole soul-spiritual significance, especially at Michaelmas. Then a person can say to themself: "You will become lord of this process, which otherwise takes its natural course outside your consciousness, if—just as you bow thankfully before the birth of the Redeemer at Christmas and experience Easter with deep inner response—you learn to experience how, at this autumn festival of Michael, there should grow in you everything that goes against love of ease, against anxiety, and makes for the unfolding of inner initiative and free, strong, courageous will." The festival of strong will—that is how we should conceive of the Michael Festival. If that is done, if nature-knowledge is true, then spiritual human self-consciousness at the Michael Festival will shine out in its true colors. Then, out of all I have described, the majestic image of Michael and the Dragon will arise once more. But this picture of Michael and the Dragon paints itself out of the cosmos. The Dragon paints himself for us, forming his body out of bluish-yellow sulfur streams. We see the Dragon shaping himself in shimmering clouds of radiance out of the sulfur-vapors; and over the Dragon rises the figure of Michael—Michael with his sword."

"But we shall picture this rightly only if we see the space where Michael displays his power and his lordship over the

Dragon as filled not with indifferent clouds but with showers of meteoric iron. These showers take form from the power that streams out from Michael's heart; they are welded together into the sword of Michael, who overcomes the Dragon with his sword of meteoric iron."

"Man can rescue himself from its impending dangers only if he starts to spiritualize life in this very domain, if he penetrates through his environment to the spiritual; if he turns from the iron which is used for making engines and looks up again to the meteoric iron which showers down from the cosmos to the earth and is the outer material from which the power of Michael is forged. Men must come to see the great significance of the following words: "Here on earth, in this epoch of materialism, you have made use of iron, in accordance with the insight gained from your observation of matter. Now, just as you must transform your vision of matter through the development of natural science into Spiritual Science, so must you rise from your former idea of iron to a perception of meteoric iron, the iron of Michael's sword. Then healing will come from what you can make of it." This is the content of the aphorism:

O Man,

You mold iron to your service,

You reveal iron

According to the value of its substance

In many of your works.

Yet it will bring you healing

Only when it reveals to you

The lofty power of its spirit.

"That is, heal oneself with the lofty power of Michael and the sword he has welded together in cosmic space out of meteoric iron. Healing will come when our material civilization proves capable of spiritualizing the power of iron into the power of Michael-iron, which gives man self-consciousness in place of mere nature-consciousness."

Supersensible Man, Rudolf Steiner, Lecture III, The Hague, November 17, 1923, GA 231

"You are, of course, familiar with the phenomenon of the sunspots which appear with a certain regularity—dark spots that are observed on the Sun. These sunspots and their meaning are the cause of much dispute in material science, but more accurate research would reveal the following. *A constant impulse arises from within the Sun to throw out Sun-substance into the Universe through these dark portals.* And the Sun-substance thus thrown out appears within our solar system in the form of *comets, meteors, and shooting stars*. Now it is particularly in our age that the Beings who rule over the Universe from within the Sun are casting forth these comets, meteors, and shooting stars. They did so in earlier times as well, but in our time, this activity of theirs has a new significance."

"You will remember, I said how in earlier times, it was the purely spiritual impulses in the starry system that were particularly at work. In our time, it is the impulses contained in the iron thrown out from the Sun that have special significance for human beings. These impulses are used by Him whom we know as the [Archangel] Michael Spirit, in the service of the spiritual in the Cosmos. In our age, there are thus present in the Cosmos impulses which were not working with the same strength in earlier periods of civilization. This cosmic iron, in its spiritual nature, makes it possible for the Michael Spirit to mediate between the super-sensible and the material on Earth.

We find therefore on the one side a spirit of warfare abroad in the world man enters, when in our time he reaches to what lies behind outer sense-existence. When a man crosses the threshold with super-sensible sight and, instead of directing his gaze to matters which concern him personally, turns his attention to great affairs of the Universe which underlie our whole civilization, then he sees warfare and battle, spiritual battle. There is strife, there is war and conflict in the spiritual worlds behind the veils of existence. And the iron which, even to the point of physical manifestation, is thrown out by the Sun Spirits into the cosmos—with this iron, Michael arms Himself for His task in the cosmic war. For Michael has the task of helping humanity to go forward in the right way in the face of these powers of strife behind the veils of civilization. On the one hand—battle and warfare; on the other hand, the labors and strivings of Michael."

"As earthly humans, we have iron in our blood. If we were beings with no iron in our blood, the feeling and impulse of freedom would still be able to arise in our souls, but we should not have bodies which could be used for putting this impulse into operation. That we are able not only to conceive the idea of freedom, but also to feel in our body the power to make the body itself into a bearer of the impulse of freedom, is due to the fact that, in our age, we can learn how Michael takes the cosmic iron, which was cast out also in former times, into His service. We ourselves, if we understand the Michael impulse aright, can learn how to place the iron that we have within us into the service of the impulse of spiritual freedom. Matter, in any case, has meaning for us only when we learn to understand it as an expression of the spiritual in the Universe. In this age, what we have to learn is to make the right use of the iron in our blood. For wherever iron is, there too is the impulse for the development of freedom. This is true in the Cosmos and true also in humanity. It was out of a deep instinct that the initiates of old ascribed iron to Mars—iron which has a significance for human blood, and therewith also a cosmic significance."

***The Evolution of the Earth and Man and The Influence of the Stars*, Rudolf Steiner, Lecture XIII, *The Nature and Task of Anthroposophy. Biela's Comet*, September 20, 1924, GA 354**

"The comet of which I've been speaking is not the only one that was merciful enough to split up and dissolve in cosmic space at the right time. There is a large number of other comets that have done the same. A great deal of superstition has always been connected with the subject of comets. Anthroposophy approaches the matter in an absolutely scientific way."

"And what has the comet done? It gives us the meteors, the fine meteor rain. Instead of colliding with the Earth and breaking people's heads, it is giving all its substance away, and this substance, every piece of it, is in the Earth. Every few years, the comet gives something to the Earth. And people who want to live by science alone and who will not admit that the Earth receives something from the Cosmos are every bit as stupid as someone who would say that when a person eats a piece of bread, it is not in him. Obviously, what the comet gives us is in the Earth, but science takes no notice of it. Where, then, is it to be found? It goes into the air, is passed from the air into the water, from the water into the roots of the plants, from the roots of the plant into the food on our tables. From there, it passes into our bodies. We eat what the comet has been giving us for centuries. This, however, has long been spiritualized. Instead of the comet putting an end to the Earth, its substance has long been in the Earth as a means of earthly nourishment, and it is a remedy, a cosmic remedy: it alleviates nervous troubles in human beings. The comets appear out there in the heavens, and after a time, they find their way into us out of the Earth. By that time, their substance has become spiritualized."

Meteors and Meteorites

The Four Seasons and the Archangels, Rudolf Steiner, Lecture I, *The Michael Imagination,* October 5, 1923GA 229

"And we shall form a true picture of autumn only if, on the one hand, up above, we have the powerful Michael-Imagination, with the sword forged from meteoric iron, the garment woven out of Sun-gold and shot through with the Earth's silver-sparkling radiance, and just as the sword of Michael emerges when we think of the meteoric iron as concentrated in the sword and blended with it."

Reading the Pictures of the Apocalypse, Rudolf Steiner, Lecture XII, Kristiania, May 21, 1909, GA 104a

"In the age represented by *the seven seals, something like a shower of meteorites will occur, caused by increasing materialism, and some human beings will ascend to a spiritual state.* What the spiritualized human beings have acquired through their efforts in our Post-Atlantean Age will completely permeate them within. When, in the Age of the Sixth Seal, everything that the human being has in terms of Sentient Soul, Intellectual Soul, and Consciousness Soul has been worked into the other members, human beings will have achieved the ability to create an external imprint of their inner life in their gesture, features, in their whole life. Because they have worked on their development, they will be able, in the Fourth, Fifth, and Sixth Ages in the Epoch of the Seals, to use these three soul forces—the Sentient, Intellectual, and Consciousness Souls—to permeate and work on themselves in order to take in Manas [Spirit-Self]."

Cosmic Workings in Earth and Man, Rudolf Steiner, Lecture IV, _Effects of Substances in the Cosmos and in the Human Body_, October 27, 1923, Dornach, GA 351

"On the other hand, all that which happens less regularly—which certainly can be calculated but still happens less regularly, for example, comets and meteors—all these phenomena are connected with what is "free will" in man, with what gives rise to free will in man. First and foremost, we must turn our attention to a substance which is particularly important, which is abundant on the Earth and indeed in the Universe and is present in meteors which fall onto the Earth."

"And iron must be there in order that we may have the power to exercise free will. Mars gives us the power of iron; meteors, since they are all the time giving up iron to the air, supply the substance of iron. Mars is that body in the Cosmos which enables us to use in the proper way that iron which the meteors and comets bring to us in an irregular manner. It is actually the force of Mars together with that of the comets and meteors which enables us to speak."

***The Four Seasons and the Archangels,* Rudolf Steiner, Lecture I, *The Michael Imagination*, Rudolf Steiner, Dornach, October 5, 1923, GA 229**

"And when, in high summer, from a particular constellation, meteors fall in great showers of cosmic iron, then this *cosmic iron*, which carries an enormously powerful healing force, *is the weapon which the gods bring to bear against Ahriman*, as dragon-like he tries to coil round the shining forms of men. The force which falls on the earth in the meteoric iron is indeed a cosmic force whereby the higher gods endeavor to gain a victory over the ahrimanic powers when autumn comes on. And this majestic display in cosmic space, when the August meteor showers stream down into the human shining in the astral light, has its counterpart—so gentle and apparently so small—in a change that occurs in the human blood. This human blood, which is in truth not such a material thing as present-day science imagines but is permeated throughout by impulses from soul and spirit, is rayed through by the force which is carried as iron into the blood and wages war there on anxiety, fear, and hate. The processes which are set going in every blood-corpuscle when the force of iron shoots into it are the same, on a minute human scale, as those which take place when meteors fall in a shining stream through the air. This permeation of human blood by the *anxiety-dispelling force of iron is a meteoric activity*. The effect of the raying in of the iron is to drive fear and anxiety out of the blood."

"And so, as the gods with their meteors wage war on the spirit who would like to radiate fear over all the earth through his coiling serpent-form, and while they cause iron to stream radiantly into this fear-tainted atmosphere, which reaches its peak when autumn approaches or when summer wanes—so the same process occurs inwardly in man, when his blood is permeated with iron. We can understand these things only if we understand their inner spiritual significance on the one hand, and if on the other, we recognize how the sulfur process and the iron-process in man are connected with corresponding events in the cosmos."

"A man who looks out into space and sees a shooting-star should say to himself, with reverence for the gods: "That occurrence in the great expanse of space has its minute counterpart continuously in myself. There are the shooting-stars, while in every one of my blood-corpuscles iron is taking form: my life is full of shooting-stars, miniature shooting-stars." And this inner fall of shooting-stars, pointing to the life of the blood, is especially important when autumn approaches, when the sulfur process is at its peak. For when men are shining like glow-worms in the way I have described, then the counterforce is present also, for millions of tiny meteors are scintillating inwardly in their blood."

"This is the connection between the inner man and the universe. And then we can see how, especially when autumn is approaching, there is a great raying-out of sulfur from the nerve-system towards the brain. The whole man can then be seen as a sulfur illuminated phantom, so to speak."

"But raying into this bluish-yellow sulfur atmosphere comes the meteor swarms from the blood. That is the other phantom. While the sulfur-phantom rises in clouds from the lower part of man towards his head, the iron-forming process rays out from his head and pours itself like a stream of meteors into the life of the blood."

"Such is man, when Michaelmas draws near. And he must learn to make conscious use of the meteoric force in his blood.

He must learn to keep the Michael Festival by making it a festival for the conquest of anxiety and fear; a festival of inner strength and initiative; a festival for the commemoration of selfless self-consciousness."

"Just as at Christmas we celebrate the birth of the Redeemer, and at Easter the death and resurrection of the Redeemer, and as at St. John's Tide we celebrate the outpouring of human souls into cosmic space, so at Michaelmas—if the Michael Festival is to be rightly understood—we must celebrate that which lives spiritually in the sulphurating and meteorizing process in man, and should stand before human consciousness in its whole soul-spiritual significance, especially at Michaelmas. Then a man can say to himself: "You will become lord of this process, which otherwise takes its natural course outside your consciousness, if—just as you bow thankfully before the birth of the Redeemer at Christmas and experience Easter with deep inner response—you learn to experience how, at this autumn festival of Michael, there should grow in you everything that goes against love of ease, against anxiety, and makes for the unfolding of inner initiative and free, strong, courageous will." The festival of strong will—that is how we should conceive of the Michael Festival. If that is done, if nature knowledge is true, spiritual human self-consciousness, then the Michael Festival will shine out in its true colors."

"But before mankind can think of celebrating the Michael Festival, there will have to be a renewal in human souls. It is the renewal of the whole soul-disposition of men that should be celebrated at the Michael Festival—not as an outward or conventional ceremony, but as a festival which renews the whole inner man."

“Then, out of all I have described, the majestic image of Michael and the Dragon will arise once more. But this picture of Michael and the Dragon paints itself out of the cosmos. The Dragon paints himself for us, forming his body out of bluish-yellow sulfur streams. We see the Dragon shaping himself in shimmering clouds of radiance out of the sulfur vapors; and over the Dragon rises the figure of Michael, Michael with his sword.”

“But we shall picture this rightly only if we see the space where Michael displays his power and his lordship over the dragon as filled not with indifferent clouds but with showers of meteoric iron. These showers take form from the power that streams out from Michael’s heart; they are welded together into the sword of Michael, who overcomes the Dragon with his sword of meteoric iron.”

“If we understand what is going on in the universe and in man, then the cosmos itself will paint from out of its own forces. Then one does not lay on this or that color according to human ideas, but one paints, in harmony with divine powers, the world which expresses their being, the whole being of Michael and the Dragon, as it can hover before one. A renewal of the old pictures comes about if one can paint out of direct contemplation of the cosmos. Then the pictures will show what is really there, and not what fanciful individuals may somehow portray in pictures of Michael and the Dragon”.

“Then men will come to understand these things, and to reflect on them with understanding, and they will bring mind and feeling and willpower to meet the autumn in the course of the year. Then, at the beginning of autumn, at the Michael Festival, the picture of Michael with the Dragon will stand there to act as a powerful summons, a powerful spur to action, which must work on men in the midst of the events of our times. And then we shall understand how this impulse points symbolically to something in which the whole destiny—perhaps indeed the tragedy—of our epoch is being played out."

"During the last three or four centuries, we have developed a magnificent natural science and a far-reaching technology, based on the most widely distributed material to be found on earth. We have learnt to make out of iron nearly all the most essential and important things produced by mankind in a materialistic age. In our locomotives, our factories, on all sides, we see how we have built up this whole material civilization on iron, or on steel, which is only iron transformed. And all the uses to which iron is put are a symbolic indication of how we have built our whole life and outlook out of matter and want to go on doing so. But that is a downward-leading path. Man can rescue himself from its impending dangers only if he starts to spiritualize life in this very domain, if he penetrates through his environment to the spiritual; if he turns from the iron which is used for making engines and looks up again to the meteoric iron which showers down from the cosmos to the earth and is the outer material from which the power of Michael is forged. Men must come to see the great significance of the following words: "Here on earth, in this epoch of materialism, you have made use of iron, in accordance with the insight gained from your observation of matter. Now, just as you must transform your vision of matter through the development of natural science into Spiritual Science, so must you rise from your former idea of iron to a perception of meteoric iron, the iron of Michael's sword. Then healing will come from what you can make of it." This is the content of the aphorism:

O Man,
You mold it (iron) to your service,
You reveal it (iron) according to the value of its substance
In many of your works.
Yet it will bring you healing
Only when it reveals to you
The lofty power of its spirit.

"That is, the lofty power of Michael, with the sword he has welded together in cosmic space out of meteoric iron. Healing will come when our material civilization proves capable of spiritualizing the power of iron into the power of Michael-iron, which gives man self-consciousness in place of mere nature-consciousness."

"You have seen that precisely the most important demand of our time, the Michael-demand, is implicit in this aphorism, this script that reveals itself in the astral light."

The Spiritual Beings in the Heavenly Bodies and in the Kingdoms of Nature, Rudolf Steiner, Lecture VII, April 10, 1912, Helsinki, GA 136

"There we have the planets themselves, with the fixed star, the Sun, as their ruler; and circling the planets, we have their moons (speaking of our Earth, we have our own Moon)—and within the planetary system, we also have those remarkable stars, which are so difficult for the external consciousness to place in the collective picture of the system—the meteoric and cometary bodies. We will, to begin with, look away from everything else in the stellar systems, and fix our attention upon the following quaternary in a planetary system: the planets, the fixed stars, the moons, and the comets."

"And in a curious way, the occult vision, which for a time has had the opportunity of observing cometary life, shows us that everything of a cometary or meteoric nature is ever striving to collect round itself all the harmful astral products, and to remove them out of the planetary system. We shall see further in the course of these lectures how this affects in particular the harmful astral products of humanity; but we see that the noxious Luciferic evils are carried away through the comets out of the planetary system."

***Lectures to the Workmen, The Human Being in Body, Soul and Spirit*, Rudolf Steiner, Dornach, Switzerland, April 10, 1920, GA 354**

"So, we find meteorites; they are the remains of shattered comets. *But the comets themselves are something quite different; they are alive.* The Sun, too, is alive, has a stomach, and not only consumes comets, but eats exactly like we do."

***The Etherization of the Blood*, Rudolf Steiner, October 1, 1911, Basel GA 130**

"Electricity is light in the sub-material state. Light is there, compressed to the utmost degree. An inward quality too must be ascribed to light; light is itself at every point in space. Warmth can expand in the three dimensions of space. In light, there is a fourth; it is of fourfold extensions—it has the quality of inwardness as a fourth dimension."

***Planetary Spheres and their Influence on Man's Life on Earth and in the Spiritual Worlds*, Rudolf Steiner, Lecture II, *The Cosmic Origin of the Human Form*, August 22, 1922, Oxford, GA 211**

"Suppose that you could take the human being as you see him here and turn him inside out so that the inside of him—the heart, for instance—would become outer surface. Physical man, you will readily believe, could not stay alive under these conditions. But if one could do this, taking hold of him in the inmost heart and turning him inside out like a glove, then man would not remain man as we see him here; he would enlarge into a Universe. For if we have the faculty to concentrate in a single point within the heart and thence to turn ourselves inside out in spirit, we simply do become the Universe which in the normal course we experience between death and a new birth. Such is the secret of the inner man. It is only in the physical world that he cannot be turned inside out. The heart of man, however, is in effect a Universe turned inside out, and that is how the physical and earthly world is really joined to the spiritual. We must get used to the 'turning inside out.' If we do not, we gain no real idea of how the physical world which surrounds us here is related to the spiritual world. These are the things which I was wanting to impart today."

***Cosmic Workings in Earth and Man—Effects of Substances in the Cosmos and in the Human Body*, Rudolf Steiner, Lecture I, *The Threefold Sun and the Risen Christ*, April 24, 1922, London GA 351**

"And iron must be there in order that we may have the power to exercise free will. Mars gives us the power of iron; meteors, since they are all the time giving up iron to the air, supply the substance of iron. Mars is that body in the cosmos which enables us to use in the proper way that iron which the meteors and comets bring to us in an irregular manner. It is actually the force of Mars together with that of the comets and meteors which enables us to speak."

Anthroposophists on Meteorites and Comets

The Reappearance of Christ in the Etheric, **Rudolf Steiner, Lecture IV,** ***Mysteries of the Universe: Comets and the Moon,*** **Stuttgart, March 5, 1910, GA 118**

"This is how the working of the comet now expresses itself: when the human "I" takes a step forward, then, in order that the whole man can advance, the physical and etheric or life bodies must be correspondingly transformed. If the "I" is to think differently in the nineteenth century from the way it thought in the eighteenth, there must also be something that changes the outer expression of the "I" in the physical and etheric bodies—and this something is the comet! *The comet works upon the physical and etheric or life bodies of man in such a way that they actually create organs, delicate organs that are suitable for the further development of the "I"*—the "I"-consciousness as it has developed especially since the imbedding of the Christ impulse in the Earth. Since that time, the significance of the comet's appearance is that the "I", as it develops from stage to stage, receives physical and etheric organs it can use."

Anthroposophy and Astronomy, by Dr. Elizabeth Vreede, *Anthroposophy a Quarterly Review of Spiritual Science*, Vol. 3, No. 3, Michaelmas, 1928

"We may pursue the same line of study in reference to other cosmic phenomena, the meteors—for they, too, are connected with comets. Comets are spiritual formations sent forth into the Cosmos by the highest, most sublime spirits, but in the course of their activity as purifiers of the astral atmosphere, they gather to themselves various substances, gases, and the like. Now, other cosmic formations, generated by the abnormal spirits from the realm of the Thrones—Beings who ought really to have become Seraphim and Cherubim but who have remained at the stage of the Thrones—fling themselves in the path of the meteors and work with a prodigious power acquired from their having remained at a lower level. The Thrones are the creators of the mineral group-souls, and the retarded (rebel) Thrones—those who have remained behind—generate solid mineral formations in the Cosmos, in the meteors which often accompany the comets or appear in 'showers' at more or less regular intervals of time in the atmosphere of the Earth. And so, in these formations—withdrawn as they are in a certain sense from the ordinary law obtained in the planetary system—we have the highest spirituality united with the densest materiality."

"The very behavior of comets may continually remind us that the processes taking place in the solar system can never be exhaustively explained by the Newtonian theory of gravitation. It is only in the comets and meteors that the 'incalculable,' the arbitrary, makes itself manifest. The incommensurable factor in the movements of the other heavenly bodies shows that we must base our observation of their movements on something other than the theory of gravitation which holds good for the Earth."

"The stars bear for us the traces of the deeds of the gods which lead through the Beings of the hierarchies to the frontiers of divinity itself. And in this sense, Rudolf Steiner was able to say to his hearers when he had led them from Anthroposophy to astronomy: "When we observe the life of the world of stars, we are beholding the bodies of the gods and, ultimately, of Divinity itself."

Introductory Essays on Rudolf Steiner's Star-Knowledge, The Cosmic Feminine, Steiner's View of Cometary Life, R. S. W. Bobbette, 1998

Comets, as noted by astronomers themselves, are a bewilderingly disparate group of objects. Comets have very long tails, or almost none; they skim the sun within 300,000 miles (where temperatures vaporize metals) or approach no closer than the cold space near earth; return every 3 to10 years, or a few dozen or hundreds, or millions of years, or perhaps never. They can have beautiful symmetry or are strangely asymmetrical.

Modern scientists are trying to accommodate comets as variations on a theme—not an easy task despite the fact that everyone knows a comet when they see one. Rudolf Steiner, in the course of his clairvoyant investigations, made regular descriptions of the nature of comets. He was not a theorist, and no "unifying concept" is intended. His observations should be added to those of material science, avoiding modern extrapolations and theories, and of course, understanding the spiritual basis of his method. For the purpose of initial surveys of Steiner's facts, I prefer to simply list pertinent references chronologically.

"His first description of comets is of a strictly esoteric character, given on September 2, 1906: "A comet is an assemblage of Kama, desire-substance, without the corresponding spiritual substance. The comet gets as far only as the astral body. The visibility of comets arises from the powerful friction caused by the astral body passing through etheric substance." (*At the Gates of Spiritual Science,* 1970, pg. 146)

"In the same year, Steiner predicted the occurrence of cyanides associated with comets due to their reflection of a previous stage of solar system evolution. The next reference I have is from March 5, 1910, and includes a reference back to this prediction, which by then had been spectroscopically demonstrated. This followed a comparative description of certain aspects of comets and moon." (*The Reappearance of Christ in the Etheric,* GA 118, 1983, pg. 57)

Is there perhaps something in our solar system that brings about, as a kind of mirror-image on earth, the contrast between man and woman? Yes, this higher polarity can be designated as the contrast between the cometary and lunar natures, between comets and the Moon. Just as the contrast of Sun-Earth is reflected in our head and limbs, so in feminine and masculine is reflected the contrast of Comet-Moon.

Then came the comment on cyanides noted above, pp. 60-61: "We can also express it this way: as what is done by a woman out of passion, out of feeling, is related to the dry, reasonable masculine judgement, so is the regular, reasonable course of the Moon related to the cometary phenomenon that projects apparently irregularly into our existence. Mark well—I do not mean the spiritual life of a woman but the feminine spiritual life. There is a difference. The spiritual life of a woman naturally includes masculine characteristics." (*The Reappearance of Christ in the Etheric*, GA 118, 1983, pp. 62-63)

"The comets appear at great intervals of time. Let us ask: when they appear, is their relation to human evolution as a whole such that they stimulate, as it were, the feminine principle in human nature? There is, for example, Halley's Comet, which now again has a certain actuality. The same could be said of many other comets. Halley's Comet has a quite definite task, and everything else that it brings with it stands in a particular connection to this task. Halley's Comet—we are speaking here of its spiritual aspect—has the task of impressing on human nature its own special being in such a way that this human nature and essence take a further step in the development of the I when the comet comes near the earth. It is that step which leads the I out to concepts on the physical plane. To begin with, the comet has its special influence on the two lower members of human nature, on what is masculine and feminine; there it joins company with the workings of the Moon. When the comet is not there, the workings of the Moon are one-sided; the workings change when the comet is present. The comet works upon the physical and etheric or life bodies of humanity in such a way that they actually create organs, delicate organs that are suitable for the further development of the I; the I-consciousness as it has developed especially since the embedding of the Christ impulse in the earth. Since that time, the significance of the comet's appearance is that the I, as it develops from stage to stage, receives physical and etheric organs it can use." (*The Reappearance of Christ in the Etheric*, GA 118, 1983, pp. 64-65)

Every cometary body thus has a definite task. Human spiritual life takes its course with a certain cosmic regularity, as it were—a bourgeois regularity one could say. In contrast to these events there are things that always bring a step forward, that are naturally distributed over wider spans of time; these events occur under the influence of the comets. The various comets have here their different tasks, and when a comet has served its purpose, it splinters. Thus, we find that from a certain point of time onward, some comets appear as two and then splinter. They dissolve when they have completed their tasks. Regularity, all that belongs to the common round, is connected with the lunar influence; the entry of an elemental impulse, always incorporating something new, is connected with the influence of the comets.

"As when a new little child is born, so it is when, through the return of a comet, something quite new is produced. We must remember, however, that with certain comets the I is always driven more and more into the physical world, and this is something we must resist. A reappearance of Halley's Comet should therefore give us warning that it might prove to be an evil guest if we were simply to give ourselves up to it, if we were not to resist its influence." (*The Reappearance of Christ in the Etheric,* GA 118, 1983, pp. 66-67)

"On the other hand, is it not wonderful that cometary existence takes hold of the depths of life, including the animal and plant life that is bound up with human life? Those who pay close enough attention to such things would observe how there is actually something altogether different in the blossoming of flowers from what is usually the case." (*The Reappearance of Christ in the Etheric,* GA 118, 1983, pg. 68)

Slightly later, on April 18, 1910, Steiner discussed these points again, more briefly and from a slightly different angle: "Just as the movements of the planets circling the Sun correspond to the regular events in the evolution of humanity, so the appearance of a comet corresponds to an influence that runs counter to the regular events. Rosicrucian research has demonstrated that every comet exerts a particular influence on human evolution. The present comet has as its particular influence an intense impulse towards materialism. Every time Halley's Comet has appeared, a new impulse toward materialism has taken place. Its appearance in 1759 corresponded with the epoch in which Voltaireanism was at its high point. The appearance in 1835 corresponded with the materialism of Moleschott, Buchner, and others. In the same way, in our time, there will be a new impulse toward materialism, and the outer sign of this is the appearance of the comet. Those who let themselves be swayed by its influence will fall into the deepest materialism." (*The Reappearance of Christ in the Etheric*, GA 118, 1983, pg. 107)

During the next month, on May 10, 1910, the1906 prediction was again referred to: "There is thus a lunar body, the Moon, which has gone too far in its evolution, and such bodies have remained at an earlier stage of evolution, namely, the comets. A comet brings with it the laws of the earlier "Moon" incarnation of the Earth and therefore renews these ancient laws. It also brings with it the cyanide compounds, as has recently been established by outer science and has been known for a long time to occult science. Just as oxygen and carbon compounds are necessary to us on earth, so the cyanides were essential on the "Moon" incarnation of the Earth." (*The Reappearance of Christ in the Etheric*, GA 118, 1983, pg. 113)

From April 3-14, 1912, Dr. Steiner gave extensive descriptions of the solar system, which included these somewhat dramatic points about comets: "From man himself stream forth continually, erroneous, base, evil thoughts; these are, as we know, realities, which pass out into the astral world and continue to exist there; so that the astral sphere of a planet is filled, not only by the normal substances of its psychic being, but also by this out-poured substance. In a curious way, the occult vision, which for a time has had the opportunity of observing cometary life, shows us that everything of a cometary or meteoric nature is ever striving to collect around itself all the harmful astral products, and remove them out of the planetary system."

"Occult vision shows us, that, however, as a matter of fact, just about when the comet disappears from physical sight, it dissolves and continues its way through a world not limited to the ordinary three dimensions of space; it no longer exists in the ordinary world. It actually disappears on the one side and appears again on the other. Thus, in short, with the exception of a few comets which really have long-drawn elliptical orbits, the comets for the most part are so formed that they come in at the one side and disappear at the other, and when they reappear, they have formed themselves anew. Why? Because as a comet approaches, it exercises a power of attraction—it is, at first, merely a sort of spiritual force center; this center attracts all the harmful astral streams, develops them around itself, and thus takes on form. In its journey towards the other side, it draws this along with it, until it gets beyond the region of the planetary system, when it casts it off into cosmic space. Then the center of force builds itself again at the other pole without needing three-dimensional space, once more takes up the harmful matter, and throws it out the other side."

"The sphere of the comets extends to the sphere of the Cherubim. We must therefore first know what is the nature of the Cherubim and Seraphim in order to understand the substance and movement of the comets." (*The Spiritual Beings in the Heavenly Bodies and the Kingdoms of Nature*, GA 136, 1951, pp. 121-130)

On January 4, 1921, Steiner began a series of comments on comets as part of a fairly extensive review of astronomy: "Upon these two thought-pictures,—that the orbital planes of the planets lie in the proximity of the plane of the Sun's equator and that the orbits are eccentric ellipses,—Kant,

Laplace, and their successors built up the nebular hypothesis. Follow what emerges from this. At a pinch, and indeed only at a pinch, it is a way of imagining the origin of the solar system. But the astronomical system thus constructed contains no satisfactory explanation of the part played by the cometary bodies. They always fall out of the theory. This discordance of the comets with the theories which were formed, as described, in the course of scientific history, proves that the cometary life somehow rebels against a concept formed, not from the whole but from a part of the whole. We must be clear too, that the paths of the comets frequently coincide with those of other bodies which also play into our system and present a riddle precisely through their association with the comets. These are the meteoric swarms, whose paths very frequently—perhaps even always—coincide with the cometary paths. Here, my dear friends, taking into account the totality of our system, we are led to say: A sum of ideas has gradually been formed from the study of our planetary system as a whole,—ideas with which we cannot do justice to the seemingly irregular and almost arbitrary paths of the comets and meteoric swarms. They simply refuse to be included in the more abstract pictures that have been reached." (*The Relation of the Diverse Branches of Natural Science to Astronomy*, GA 323, pp. IV. 7-8)

And then on January 8 he says: "So for the comets: by spectroscopic investigation and by comparison with the corresponding results for the planets, the idea arose that the phenomena might be explained electromagnetically. Such ideas will at most lead to analogies, which may no doubt have some connection with reality, but which will certainly not satisfy us if we are looking into it more deeply."

"Yet as I said, leaving this aside, there was one thing which emerged quite inevitably when the phenomena of comets were studied in more detail. While for the rest of the planetary system they always speak of gravitational forces, the peculiar position of the comet's tail in relation to the Sun inevitably drove the scientists to speak of forces of repulsion from the Sun—forces, as it were, of recoil. The terminology is not the main point; it will of course vary with the prevailing fashion. The point is that science was here obliged to look for something in addition to—and indeed opposite to—gravity. In effect, with the comets, something different enters our planetary system,—something which in its nature is in a way opposite to the inner structure of the planetary system as such."

"...Yet in the early 19th century, they had statistics purporting to connect them (comets) with good and bad vintage years. These too, occur somewhat irregularly; their sequence does not seem to follow regular laws of nature. And even Hegel did not quite escape this conclusion. He thought it plausible that the appearance or non-appearance of comets should have to do with the good and bad vintage years."

"...Kepler said: In the great Universe—even the Universe into which we look by night—there are as many comets as there are fishes in the ocean. We only see very, very few among them, while all the rest remain invisible either because they are too small or for some other reason. Even external research has tended to confirm Kepler's saying. The comets seen were recorded in olden times, and it is possible to compare the number. Since the invention of the telescope, ever so many have been seen than before. Also, when looking out into the starry sky under different conditions of illumination—that is to say, making provision for extreme darkness—a larger number of comets are recorded than otherwise. Even empirical research therefore comes near to what Kepler exclaimed, inspired as he was by a deep feeling for Nature."

"Now if one speaks at all of a connection between the Cosmos and what happens on the Earth, it surely is not right to dwell one-sidedly on the relation to our Earth of the other planets of our system and to omit heavenly bodies which come and go as the comets do. It is especially one-sided since we must now admit that comets give rise to phenomena indicating the presence of quite other forces—forces opposite in kind to those which we usually attribute the cohesion of our planetary system. The comets do in fact bear something opposite into our system, and if we follow it up, we must admit that this too is of great significance. Something in some way opposite in nature to the force which holds it together comes with the comets into our planetary system."

"Now that we have ascertained all this about the cometary bodies, let me compare the relation between our planetary system and the comets to what is there in the ovum of the female germ-cell, in its relation to the male element, the fertilizing sperm. Try to imagine, try to visualize the two processes, as you might actually see them. There is the planetary system; it receives something new into itself, namely the effects of a comet. There is the ovum, it receives into itself the fertilizing effect of the male cell, the spermatic zoid."

"Looking at the two phenomena side by side without prejudice, as you might do in ordinary life when you see two things, obviously comparable, side by side. Do you not find plenty of comparable features when you contemplate these two? I do not mean to set up any theory or hypothesis. I only want to indicate what you will see for yourselves if you once look at these things in their true connection." (*The Relationship of the Diverse Branches of Natural Science to Astronomy*, GA 323, pp. VIII, 114-19)

At the end of these particular lectures, January 18, 1921, a rather exceptional perspective on comets was given: "Truth is that in studying cometary phenomena we get into difficulties if we conceive the cometary body too in the same way we are wont to think of a planetary body. The planetary body—the planetary body you may represent as though it were a self-contained body moving through space. You will not go much against the facts in so conceiving it. Not so a cometary body. Again and again, you will find yourself in contradiction to the phenomena if you conceive it after the same pattern as the planetary body. You will never understand the cometary body, in the way it moves—or seems to move—through cosmic space, if you regard it as you are accustomed to regard the planetary bodies."

"See what becomes of it on the other hand if you regard it as I shall now describe. Take all the empirical facts that are available and try to thread them on this line of thought. Imagine that in this direction—towards the Sun, as we may say—the comet comes into being at every moment. It is forever coming into existence in this direction. It pushes forward its cometary nucleus, or what appears as such. Behind, it melts away again. In this way it thrusts forward—forever coming into being on the one hand, passing away again on the other. It is not a body in the same sense that a planet is,—not at all. It is perpetually coming into being and passing away again—renewed in front, accruing all the time in this direction, losing the old in its tail. It pushes forward like a mere effulgence, a mere phenomenon of light; but please, I do not say that that is all it is."

"And now remember what we were saying a few days ago. There is not merely the Moon up there and the Earth here, but every planet has a certain sphere, and what we see is only a point at the periphery of the said sphere. The true Moon is the sphere bounded by the lunar orbit. We, with the Earth, are in the lunar sphere. So also, in a certain sense, we are in the Solar Sphere and in the spheres of all the planets. The planets are not merely what is out there, moving in lemniscates,—what is at yonder point or yonder at any given moment. The visible point is only a specialized part of the whole; it is, as I was saying, like the area of germination in the germinal vesicle of the human embryo."

"If you remember this, then you will say to yourselves: Here now I have the Earth and Sun. In fact, two spheres are interpenetrating, thrusting into each other,—spheres which are really due to materialities of opposite tendency and kind. The one comes from the center of the Sun, towards which negative matter is tending; the other from the center of the Earth, from which positive matter is raying out. Positive and negative materialities are interpenetrating here. Naturally, the interpenetration will not everywhere be homogeneous. Not even clouds that move through one another would interpenetrate homogeneously. It is essentially inhomogeneous. Imagine how, in this mutual penetration, the different densities will impinge on one another. Then, in the penetration of the one substantiality by the other, you have the requisite conditions for such phenomena as comets to arise. Comets are ever-nascent phenomena, perpetually coming into being, passing away again; and if we draw our ideal picture of a planetary system, we have not to imagine that the comet is arriving there from some great distance and then making its departure. Out there—outside the system—we need not imagine it to exist at all. It is not there to begin with, but becomes; then, at the perihelion, changes the gesture of its form, which is in fact ever-becoming, ever-nascent. Out there at last it melts away again and it is no more. The comet comes into being and passes away; that is its very nature. Hence it can sometimes have apparent paths that are not closed at all—parabolic paths or hyperbolic, —for there is nothing moving round such as would have to move in a closed path. All that there is, comes into being and may well do so in a parabolic direction and then vanish and be no more."

"Altogether, we must look upon the comet as a fleeting thing. In relation to Sun and Earth, it is a phenomena of compensation between ponderable and imponderable matter,—a meeting of the two kinds of matter, which do not immediately balance-out as when light extends in air. For in the latter instance too, there is a meeting of the ponderable and the imponderable; here however they spread continuously, homogeneously as it were, do not impinge on one another. Take for example air, with light of a certain intensity passing through it. The light spreads homogeneously; but if it so be that the light does not adapt itself to the air quickly enough, a kind of inner friction will ensue between the ponderable and imponderable matter; only I beg you not to understand this in a mechanical sense but as an inward process. Follow the comet and its movement. It is a mutual friction of ponderable and imponderable matter that moves through space. It comes into being at every moment and passes away again." (*The Relation of the Diverse Branches of Natural Science to Astronomy*, pp. XVIII, GA 323, 9-12)

Recall that I had to introduce this essay with the fact that even modern astronomers consider comets to be bewilderingly disparate and now consider what Steiner felt compelled to say about them in September 1922: "Julius Robert Mayer, however, asked this question. He found that every year a certain number of celestial bodies that are like comets fly into the sun. You see, those are the food the sun takes in. Even to this day, when we look up to the sun, we can see that the sun has a healthy appetite; it consumes a large number of comets every year. Just as we eat and thus develop energy, so the Sun eats comets, so to speak, and thus develops warmth."

"Now, gentlemen, when comets have shattered and fallen down, we find their hard iron cores. But these are only the parts of the comet that fall down. We human beings also have iron in our blood. Since comets disintegrate into meteorites, which consist of iron, people say that the comets themselves are made of iron. But this is the same nonsense as believing that the human beings consist of iron because they have iron in their blood and a small lump of iron would be all that's left after they disintegrate."

"So, we find meteorites; they are the remains of shattered comets. But the comets themselves are something quite different; they are alive. The sun, too, is alive, has a stomach, and not only consumes comets, but eats exactly like we do." (*The Human Being in Body, Soul and Spirit*, GA 9, 1989, pp. 172-173)

On July 7, 1924, Steiner repeated his point of the evolutionary 'youth' of the comets requiring that they have prussic acid, as the Earth did at an earlier stage (Old Moon). (*The Evolution of the Earth and Man*, GA 354, 1987, pp. 45-46).

Later, on September 13, he repeated the relationship of comets to the quality of vintage, in company with sunspots." (*The Evolution of the Earth and Man*, GA 354, 1987, pp. 171-173)

Although it is not exactly a description of comets, during the 1921 lectures, while discussing the nature of orbits in a mathematical sense, Dr. Steiner showed the mathematical basis for understanding his provocative observation that comets 'leave space' and reappear again at the other side. After showing the progressive transformations of the mathematical equation called the Curve of Cassini, Steiner sums up with: "So you see, mathematics itself provides us with facts which oblige us to go out of space if we would preserve the continuity of the idea. The reality itself demands of us that in our ideas we go out of space. Even in mathematics, therefore we are confronted with something which shows us that in some way we must leave space behind if the pure idea is to follow its right path. Having ourselves set going the idea in the beginning to think the process through, we must go on thinking in such a way that space is no longer of any help to us. If this were not so, we should not be able to calculate all possibilities of the equation." (*The Relation of the Diverse Branches of Natural Science to Astronomy,* GA 323, January 9, pp. IX 6-7)

The Healing Gift of Iron, Avatar Bertram von Zabern MD

The red pigment of the blood, hemoglobin, contains iron, the chemical properties of which are at work in the functions of respiration, the absorption of oxygen in the lungs, and its transport to all body tissues. An iron deficiency could cause not only fatigue and physical weakness, but in general, it could be responsible for a person's constant lack of resistance to infections. Interestingly, a decrease of iron in the blood may be itself caused by infections, thus leading to a vicious cycle of poor health.

Since the arrival of AIDS, the lack of resistance to infections is the greatest challenge of modern medicine. Immunology, like other conventional sciences, has become a research field of details resulting in ever more immunizations rather than addressing health in a holistic way. Within the scope of this article, I will refer to basic observations, which link the outer appearance of human nature, illness, and remedial substance to their inner origins. Such views have been renewed for modern thinking by Rudolf Steiner, the inaugurator of Anthroposophy, although we find this wisdom described throughout history.

Iron is the most prevalent of the heavy metals in the human body as well as in our planet as far as earth has been explored. According to W. Pelikan, a well-known anthroposophical researcher, the distribution of mineral iron stretches from North America, England, France, Germany, into Russia and even North China as a vast belt around much of the northern temperate earth latitude. Characteristically, it occurs next to a similar belt of coal formation, an indication that in the early stages of earthly evolution iron, like coal, was part of organic life. A special form of iron is found on all continents: meteorites fall from the skies to the earth every year. This meteoric iron is a cosmic pure metal with a crystalline pattern not seen in earthly iron.

Civilization has been called the "Iron Age" since humans learned to use this metal for the purposes of technology. As iron is one of the main ingredients of nature and human life, we are not surprised to find, through many epochs of history, that cultures have described their central spiritual values in connection with this metal. Never have these values been more deeply expressed than through the personality of the apostle Paul, who is always depicted with his sword. The sharp iron blade points to Paul's own transformation from a cruel, merciless persecutor to become the "instrument of the Lord Christ." Paul himself used, at the end of his epistle to the Ephesians, the picture of the armor of God to fight the hierarchic powers of evil. "The spiritual sword is the Word of God." Another sword-bearing figure in Christian depictions is the archangel Michael, about whom we read in the Apocalypse that he defeated the satanic dragon. But Christian art shows Michael also holding a scale. There, the blade of his sword is transformed into the "tongue" of the scale, pointing out the difference between good and evil.

Are these considerations relevant to understanding the immune system? Whichever way the different white blood cells and other elements of the "body defenses" work, the emotional and spiritual condition of the patient makes a vital difference. A strong sense of determination not to become ill, an attitude of guardedness against infection, are effective helpers to prevent communicable illness. Our immune system is even capable of fighting cancer with amazing results, especially in patients who do not give up.

Observations of this kind remind us of the inner strengths of iron: courage and willpower to fight. At the same time, they raise questions about the purpose of the immune system. Is it there to fight infections or cancer, to kill bacteria, viruses, or malignant cells—in terms of bacteriology, that would define the immune system. But it does not define health, because it does not include the personal experience of going through an illness to regain new health. When we are "run down," "under stress," or sleeping or eating poorly, not only our resistance to various illnesses is compromised, but we become victims of our own poor lifestyle. As disruptive as an illness can be for the job we are trying to fulfill, it helps us to take a break from an exaggerated work commitment and to gain a better perspective on life. Recovery then can bring a new resolution, a new sense of inner balance and joy.

Samuel Hahnemann, the father of homeopathy, had insight into the true process of healing that always included the mind as well as the body. He saw a deep spiritual connection between an illness and a remedy. The inner dynamic of a substance does not become active in its crude material form. To become a remedy, a substance has to be rhythmically diluted. Hahnemann taught in his book Organon that in the homeopathic process, the inner force of a natural substance is set free so it can truly heal. The understanding of the spiritual side of healing was more common in the earlier days of medicine. The Finnish epic *Kalevala* describes the power of iron to injure and kill; but the hero Wainamoinen knows the mythic origin of the iron, from where the spiritual power comes to stop the bleeding.

The cosmic side of iron is traditionally connected with Mars, the god of warfare, and his red planet. Paracelsus wrote that iron is a universal force present in the planet Mars, in the metal, in the function of the gall bladder and the bile, and in plants permeated by the force of iron, such as the stinging nettle.

As a medication to boost the body defenses in flu and similar viral infections, we use iron phosphate in its homeopathic form. The phosphorus component addresses the inflammatory nature of these illnesses. Underlying dispositions like fatigue, poor resistance, anxiety, we treat by homeopathically using meteoric iron. Supplementation of material iron has little to do with these medications, even though there are conditions, such as iron deficiency anemia, where we have to supplement nutritional iron.

Typical "iron plants," in the sense in which Paracelsus described the stinging nettle, bear the signature of iron regardless of their material iron content. One of these plants is Boneset, Eupatorium perfoliatum, a potent flu-fighting remedy. It is a heritage of native American medicine, as it only grows wild in America. We prescribe it as the Weleda flu preparation Infludo, which contains Boneset as a main ingredient, and in combination with it Ferrum phosphoricum 6X. You may find Boneset in your back yard or alongside a country road. Its leaves are serrated, lanceolated and sharply pointed. Their flavor is of a penetrating bitterness that truly wakes us up. It is a healthy bitter taste that activates bile excretion. We can see here already a glimpse of how Boneset works its magic; there is a cleansing, "clarifying" effect to that bitterness and its bile activation, which identifies it as part of the iron force described above. The name Boneset indicates that it helps with fevers generating violent bone aches. "Perfoliatum" means "through the leaf"—opposite leaves are uniquely grown together at their base, so that the stem pierces the double leaf. This martial appearance is accentuated in the young plant by the small top pair of leaves pointing upwards like the tip of a spear.

There are other plants carrying the force of iron: Echinacea, Celandine (Chelidonium), Prunus spinosa (a wild-growing bush of the plum family), and certainly the oak tree, just to name a few. If we considered them only to be fighters against infection or other illness, they would be misunderstood. They are fighting to bring about balance in our lives. Their message is: "Find yourself in your inner equilibrium, then you will understand us, and you will be healed.

Rudolf Steiner's Homeopathic Medicine, Ferrum Sidereum

The use of Meteoric iron, Ferrum sidereum, in potentized form is a discovery of Anthroposophical medicine. As the name implies, the substance originates in the cosmos and is constituted of an alloy of iron, nickel, and small amounts of cobalt, graphite and phosphorous. According to Rudolf Steiner, meteors are being sent from the Sun to the Earth and they correspond in their healing characteristic to both the physical and spiritual healing forces of the iron in the red blood cells of the human body. In the biosphere around the earth, they appear as a striking light streak. Polished surfaces of meteors are recognized by crisscrossing straight lined so-called Thornton structures.

CERN: Biggest Science Scam in History

Once you begin to understand how false flags are perpetuated on citizens of the world, whether they are local incidents like alleged shootings or false flags of global implications like 9-11, you can uncover any false flag, no matter how small or large.

The United States government made it legal to propagandize its citizenry through the expanded powers of the National Defense Authorization Act of 2012. It is no longer treasonous to lie and trick Americans into believing anything that the government wants in order to carry out its ultimate plans. Whether the agendas are centered on gun control, war, or banking bailouts, false flags use legal methods of deception to persuade the dreaming masses that these agendas are necessary for their protection and National Security.

The false flag that we want to point out to those with 'eyes to see' and 'ears to hear' is the international false flag called

CERN (European Organization for Nuclear Research) or the Large Hadron Collider. Once the curtain is lifted on this illusion, your fears that their crazy experiments will explode the entire world into a dark-matter abyss will be allayed. The truth is, the main work being done at CERN is related to the World Wide Web and has little to do with their stated mission. By seeing the full scope of the project, you can also start demanding more transparency for a so-called science operation that is stealing trillions of dollars from citizens around the world.

As Dr. Amit Goswami, a leading quantum physicist has said: "No one knows what the thing does—no one does. Firing particles at each other at the speed of light can't end well. We (physicists) have no idea what we're doing."

Peeling Back the Veils of the CERN Illusion

Just as the first atom smashers were used to create bombs, everything done at CERN is for military purposes. As is usually the case, any technology that can be weaponized for the globalist's control of the world, whether it is social media, big agriculture, energy production, or the Internet, will be used for the purpose of enslaving humanity. Make no mistake: CERN is a super hub for the global military-industrial complex.

Of course, we are not told this. A general Internet search describes CERN as a very important scientific project that is in search of creation's smallest particle, the God particle. But we also see strange religious messages in the mix, such as the statue of Shiva outside of the CERN headquarters or bizarre videos of workers dancing the Hindu Shiva Dance of Destruction.

CERN Workers Perform Dance of Destruction

Untold billions (perhaps trillions) have been poured into the biggest system of machines (so we are told) ever made, relying on a science of theoretical physics that claims it can use human machines to 'create' as God did during the first moments of creation. But you will be hard-pressed to find any useful discoveries at CERN, even after decades of research and experimentation and boatloads of international funding.

Indeed, CERN has been in operation since 1954 and has yet to create anything useful or get a single step closer to its mission of proving the Big Bang Theory. Of the many supposed projects and experiments underway right now, not one of them will lead to anything but theories that attempt to place scientists on the same level as God, or perhaps higher.

CERN is the Biggest Science Fraud in History

No one has been able to stop or thoroughly examine CERN experiments as it is sanctioned by a mysterious assembly of international associations where the ultimate authority of the project has been hidden through Jesuit-like compartmentalization that you cannot find any one leader, CEO, or entity in charge. In fact, many scientists who are employed by CERN probably don't know that it is a scam. They are just conducting their limited scope of work, which could be legitimate, given to them by a supervisor that is also working within constraints and boundaries. Everything is on a need-to-know basis.

In a process known as diffusion of responsibility, no one is ultimately responsible. In CERN's case, this means that science can do whatever it wants, whether it is creating light-speed atomic collisions, capturing mysterious "particles" that can create supernovas or black holes, genetic modification of plants, animals, and humans, or just the standard militaristic Warlord Banker's favorite—building new bombs for both sides of the conflict. CERN gives scientists permission to have free rein to rip apart or explode anything they want because they are trying to find the God particle.

The CERN scientific community makes the braggadocious statement that "Science can do what God does, or better!" With its community of 25,000 international scientists, scattered throughout the world, who are not responsible to any national law or regulation, diffusion of responsibility gives this group exclusive power and immunity to decide which experiments are worthy of its endeavors while protecting its scientists from any personal or group consequences. It also hides the real purpose of CERN: It is the New World Order super-hub of the military-industrial complex controlling the World Wide Web.

Why is CERN Located in Switzerland?

It doesn't take a rocket scientist to figure out that if you want to build something secret and keep it secret, you might consider Switzerland, where you can hide money, politics, and, in this case, the biggest machine in the world. Do we really believe that the Swiss, known only for manufacturing Swiss watches and knives, built the most complicated machine in the world?

If what you want is concealment to launder and hide stolen money, Geneva, home of internationalism and globalism (IMF, BIS, League of Nations, World Council of Churches, Swiss banks, etc.) is a perfect place to hide this scientific-financial-military hoax of all times.

Let's Review Some Key Points

The Money—No one knows who all gives to CERN or how much money has been spent to date. If you think auditing the US Federal Reserve is difficult to achieve, try auditing this international world wide web of lies! We know that the Rothschilds, Rockefellers, Warburgs, and other warlord bankers privately donate huge sums that do not appear on CERN-published budgets. There are confusing stories, multiple budgets, hidden assets, and donations that are shared between seven different cities in France and Switzerland. There are also multiple supporting corporations and associations around CERN. One public budget alone shows over a billion a year is spent, and nothing is produced except claims and computer simulations. The total amount donated to CERN has estimates that begin at $11 billion and range to $3 trillion.

You see why CERN is located in a secretive Swiss banking community? Can you imagine what it would take to get a full accounting of anything from CERN, whether financial or operational? Can you see how this provides the ultimate cloak of secrecy to the world's biggest false flag that promotes the silly Big Bang Theory?

One thing is for sure. Even though America donates the largest amount of yearly funding, all member countries share data that ensures the international war machine has plenty of weaponized technology for all sides. In this way, perpetual war, spearheaded by the banking warlords, is guaranteed, and they all continue to make money on war, no matter who is fighting.

Countries involved in CERN research include: Austria, Belgium, Bulgaria, Czech Republic, Denmark, Finland, France, Germany, Greece, Hungary, Israel, Italy, Netherlands, Norway, Poland, Portugal, Slovakia, Spain, Sweden,

Switzerland, United Kingdom, Romania, Serbia, Pakistan, Turkey, India, Japan, Russian Federation, United States of America, and the European Union. Although the U.S. funds more money for CERN than any other country, any top-secret weapon research and development is not U.S. proprietary. It is shared with U.S. enemies around the world, and U.S. taxpayers are paying to support international military-industrial warmongers.

The Politics—The Lombard bankers who financed the crusades, the East India Company, the Dutch East India Company, the British East Indian Company, and the central banks of Italy, Germany, Holland, and England are centered in Geneva. Members of these old banking families have led CERN since its inception. Even though 95% of the supposed CERN tunnels are in France, Geneva is in full control of CERN finances. A complete audit of CERN is not possible because of secretive Swiss banking laws and regulations.

The Machine—All other atom smashers in the world are either above ground or have continuous ground level access points around the circle or into the tunnels. CERN is conveniently 'hidden' 575 feet underground, in tunnels that did not upset the landscape while being built, or while currently in operation. Where were the environmentalists and concerned citizens while CERN built this underground behemoth? Where are the pictures of the vast amounts of dirt that had to be excavated? Frankly, when we lived in Boston during the 'big dig', it was impossible not to see traffic congestion, cost overruns, corruption, and lots of trucks, construction workers, and dirt. We simply can't find anything on the Internet regarding the construction of tunnels at CERN, which is by far a bigger project than anything of its kind, ever.

There are periodic access points at seven locations for the CERN 'machine.' Each location has an elevator that goes down to the underground facilities where large machines connect to the supposed 17-mile-long rings on three different levels. But where are the construction pictures showing the tunnel excavation machines? It would be impossible to dig the tunnels without large construction equipment being seen in the access areas.

When you look at a picture of a 'normal' accelerator, it is above ground and has continuous access points. There is no reason to place an atom smasher underground, and many reasons not to do so.

What we are shown on the Internet and on CERN's website are pictures of the so-called underground chambers where different large machines have been built. But since we can't find any evidence that anything was actually built underground, we have to question the huge and costly experiments taking place at CERN, which, as you will read later, have few tangible results.

And now we learn that CERN wishes to enlarge its facility by creating an atom smasher ten times bigger than the current ones. Switzerland will be the exclusive recipient of the funds that will be ten times greater than the already existing CERN costs, which are the most expensive scientific experiments in history.

The new, ten-times-bigger ring will be tunneled under Lake Geneva and the surrounding area, one of the most expensive and exclusive areas in Europe, and yet no one has objected to living on top of the most radical and dangerous science experiments in history, not to mention the environmental issues in excavating the beautiful Swiss countryside. Perhaps the locals know that CERN is a hoax, that they are not in danger, and that they are being compensated for their silent participation. Much like the paid crisis actors that we find in false flags events.

What kind of environmental damage will happen when the enlarged tunnels are excavated under lakes, towns, and beautiful countryside? No one is asking.

CERN Data is Flawed According to the NSA

CERN claims to have found the Higgs Boson particle called the God Particle. This data is now questionable and seems to be propaganda based upon an inadequate theory. This coincides with the charges made by the National Security Agency and researchers at other accelerators that CERN data does not fit in with the standard models. It appears that CERN data is simply computer simulations of fake experiments that do not match the data from other legitimate accelerators around the world.

In an article entitled: *NSA Concerned: CERN's Data the Only Ones They Cannot Make Sense Of*, from February 4, 2014, we read that: "NSA officials are increasingly concerned about the fact that they cannot understand the data produced by the Large Hadron Collider at the European Organization for Nuclear Research CERN. We can't see any of those W-bosons, Higgs-bosons, or top quarks these folks claim to have detected. Instead, all this appears to be an unintelligible mess that shows up when one crashes highly energetic particles randomly."

According to rumors, NSA director Keith B. Alexander is very upset about "particle geeks with a damn high income" that are "unable to explain the meaning of their business to anybody else.'

CERN management stated that: "The entire physics community trusts the results because an unprecedented number of experts believe them. We are astonished that the NSA is so distrustful." Via unofficial channels, CERN assured that none of the particle physics experiments have developed anything useful, let alone anything harmful, in the past few decades.

Is CERN a False Flag Operation?

What is so astonishing is that CERN tells you on its own website what you need to look for in order to discern an Internet hoax. (Since CERN is the creator of the Internet, World Wide Web, which you can read further down the article—guess they needed to alert you when you might be caught up in an Internet fraud.)

CERN indicates three things that will tip you off to an Internet fraud or hoax:

- Technical sounding language.
- Credibility by association.
- Life or death questions concerning a human.

As you can see, CERN meets these criteria exactly. What project is more technical than CERN? It continuously creates new theoretical particles that no one, even the NASA can substantiate.

CERN is the largest scientific collaboration in the world. Therefore, its credibility by association is unparalleled with 25,000 top scientists working for it. That many scientists can't be wrong, says the new alternative method of scientist research called "scientific consensus" or "settled science."

What could be more life-threatening than opening black holes that could swallow the earth or finding anti-matter that destroys the entire Earth in the process?

Therefore, CERN, by its own definition, is a hoax, a hoax of mega-proportions. And it meets all the criteria for the twelve basic strategies of perception management.

The True Mission of CERN—Better Bombs

For the first time in history, mad scientists have built a machine/weapon that, they claim, is capable of destroying the planet. Hadron colliders, which are actually real mechanical devices, evolved during the Cold War to create small nuclear explosions which could be used by the military as a sample of future nuclear bombs. After the Cold War ended, Russia and America put an end to the astronomical costs of those machine-weapons, but Europe, with the new marketing of "peaceful use" took the industry of accelerators a step further. The result, they claim, is a seven teravolt, superconductive, superfluid "quark cannon" which purports to mass together the densest, most attractive substance of the universe, quarks, to explore the formation of quark-gluon liquids. These liquids are the explosives responsible for cosmic annihilations, such as Novas, super-Novas, and perhaps the hypothetical Big Bang of the universe. Because the Swiss device is supposed to be so large, it has the name LARGE Hadron Collider (LHC).

Even though we have not found sufficient evidence that these experiments are actually possible, or that CERN is actually doing them, we are concerned about the lack of integrity and the carelessness that a body of international physicists displayed when they conducted such dangerous experiments that could have dire consequences for humanity. No approval by the United States Congress or any other nation, or their citizenry were ever sought. It seems that there is a scheme to make the LHC a machine that is 'too big to fail,' defended by technocrats, nuclear industries, physicists, and the corporate press, with the same zeal they defended nuclear weapons during the Cold War, and big banks in current times.

CERN claims that the LHC has enough potency to create strangelets, a strange liquid explosive that is responsible for the Ice-9 reaction that converts stars into super-Novas with a device called Castor, a strangelet detector which was built to study them. The Castor Team affirms the LHC will likely create stable strangelets. CERN will shoot bullets at the speed of light at lead pellets in this superconductive, superfluid cannon. According to the most advanced theoretical research on the subject, they can create strangelets that would sink the Earth into a rock of a few kilometers in diameter. Once upon a time, this strange liquid might have been called "snake oil."

If all of this were to be believed, CERN would be developing the most powerful weapon ever known to man and many lesser, but dangerous, weapons could arise from these experiments. Thank goodness we are now aware that these mad scientists are simply modern-day snake oil salesmen participating in one of the largest money laundering schemes in history.

Refuting the Big Bang Theory

Everything done at CERN hangs on a "belief" in the Big Bang Theory of creation. The Big Bang, much like the Santa Claus hypothesis, no longer makes testable predictions wherein proponents agree that a failure would falsify the hypothesis. Instead, the theory is continually amended to account for all new, unexpected discoveries. Indeed, many young scientists now think of this as a normal process in science, just as consensus has replaced the scientific method.

In the time of Galileo and Isaac Newton, anyone who tried to say the Earth was not flat or that the Sun did not rotate around the Earth was not only called a "conspiracy theorist," but was jailed and even executed. It's hardly much different now. Today, the Big Bang Theory is pushed so vehemently, especially among higher education, that anyone who questions its validity, even in light of the countless flaws of the theory, can be subject to intense criticism, professionally and socially ostracized and rejected, given failing grades, and expelled from universities. The questioning scientist can have his/her professional licenses revoked, professional associations withheld, job terminated, and reputation ruined. All because the inquiring scientist went against the prevailing notions (scientism's dogma and doctrines).

It's no different from the times when people went against the idea that the Earth was flat. It's not about what is true. It's about what the educational and governing authorities say they want you to believe and say is true—recite the catechism. As a result, almost all professors and scientists are too afraid of being ostracized from their academic communities and losing their jobs to speak out against the preposterous, temporary theories of settled science.

If you want to believe in the Big Bang Theory, you must believe in it in one of three ways: by faith, by ignorance, or by indoctrination (submission). By faith, because you can't believe something which does not have adequate scientific evidence and is only a philosophical viewpoint. By ignorance, because the only way to be certain in your mind that the Big Bang Theory could work is having the facts; but, you are in a state of total ignorance and don't have facts, just beliefs told to you by other believers. Or by indoctrination, because you haven't made a logical conclusion with your own rational mind, you simply never questioned what you were told.

You must decide for yourself what you will believe because humans weren't around during the Big Bang. If you decide to believe in the Big Bang Theory, just realize that since it is not supported by science you must simply believe it as a philosophical viewpoint, not as a scientific fact.

Yet the foundational science of CERN, and all of the trillions that have gone into supporting this dark behemoth for decades, is based upon the impossible theory of the Big Bang.

Ten Basic Reasons the Big Bang Theory is Impossible

Granted, we are not trained physicists or highly paid university researchers. We are regular citizens, with average intelligence, using basic high school science principles to understand what CERN claims to do. And if we see this many holes in their highfalutin science, then why aren't they seeing it? Or better yet, we challenge any scientist to refute these points.

Here are the questions we would like some of these 25,000 highly educated scientists to clear up for us before we send you another dollar:

- If the big bang theory were true, a magnetic monopole should be one of the most prevalent (common) particles in the universe. However, instead it is the complete opposite—a magnetic monopole has never been observed.

- The curvature of matter and energy in regard to density remains very small so the probability that a Big Bang could have occurred to create the current universe is so astronomically slim that it is entirely improbable.

- If the Big Bang occurred 13.7 billion years ago, then the deeper we peer into the universe, the closer to the Big

Bang, then we should see the origin point. However, no matter how deep we peer into space, still we see no evidence of a Big Bang.

- For a Big Bang to have occurred, galaxies would be perfectly evenly spread out. Thus, the lack of universal galactic uniformity contradicts the fundamental aspects of the Big Bang Theory.

- Dark Matter and Dark Energy have never been proven, or observed in any way whatsoever, yet the Big Bang Theory depends on the existence of such potentially mythological substances. In order for the Big Bang Theory to even be valid, Dark Matter and Dark Energy would have to be the most abundant things in the universe.

- Big Bang theorists have tried to use a magical effect called "Inflation" to solve several of the obvious problems, including the Horizon Problem and the Flatness Problem. Inflation states that after the Big Bang, all the particles in the universe traveled faster than the speed of light. But Einstein's General Law of Relativity proves that nothing can travel faster than the speed of light.

- For the Big Bang to have happened and created the whole universe as we know it, the opposite thing would have happened: all matter would have moved toward order. This is impossible.

- Newton's First Law of Thermodynamics states generally that matter cannot be created nor destroyed. This is a fundamental law of science which says that matter can only be converted into other matter. Proponents of the Big Bang Theory try to say that the universe was created out of nothing. Obviously, this is scientifically impossible.

- The Big Bang Theory opposes Occam's Razor because it can only exist with innumerable adjustable parameters.
- The universe is too big to have formed in only 10-20 billion years as the Big Bang Theory suggests, since the Big Bang is theorized to have happened only about 13.7 billion years ago. This is because the speed of matter is limited by the speed of light. The problem here is that if the Big Bang had occurred, firstly the universe is too large to have only happened 13.7 billion years ago, and secondly there is temperature uniformity which requires matter to have moved beyond the speed of light to become universally uniform. This of course, is impossible according to Einstein's General Theory of Relativity, because nothing can move faster than the speed of light.
- The addition of three more theorical aspects of the Big Bang Theory have been added due to the theory being entirely inadequate: Stars III, Stars II, and Stars I. It is a changing theory that does not work.

Perhaps never in the history of science has so much quality evidence accumulated against a model so widely accepted within a field. If our limited high school science background can question this, why aren't more university professors calling out the HOAX.

Even the most basic elements of the theory, the expansion of the universe and the fireball remnant radiation, remain interpretations with credible alternative explanations. One must wonder why, in this circumstance, that four good alternative models are not even being comparatively discussed by most astronomers. These four alternatives are:

1. The Steady-State model of the universe suggests the universe always had and will always have the same density. The theory reconciles the apparent evidence that the universe is expanding by suggesting that the universe generates matter at a rate proportionate to the universe's rate of expansion.

2. The Ekpyrotic model suggests our universe is the result of a collision of two three-dimensional worlds on a hidden fourth dimension. It doesn't conflict with the Big Bang theory completely, as after a certain amount of time it aligns with the events described in the Big Bang Theory.

3. The Big Bounce Theory suggests our universe is one of a series of universes that first expand, then contract again. The cycle repeats after several billion years.

4. Plasma Cosmology attempts to describe the universe in terms of the electrodynamic properties of the universe. Plasma is an ionized gas, which means it's a gas with free-roaming electrons that can conduct electricity.

CERN Did Not Discover the Higgs Boson Particle

In an article by Jim Algar of *Tech Times* we read that: "The elusive Higgs boson, the so-called "God particle", may not have been discovered despite claims of it being detected, some scientists are saying. Particle physicists at the European Organization for Nuclear Research (CERN) announced two years ago they had discovered the Higgs particle, considered the foundation particle in the Standard Model of Particle Physics, and a Nobel Prize was awarded to Peter Higgs and

Francois Englert for their work on the theory of the Higgs Boson."

"Now, though, researchers at the University of Southern Denmark's Center for Cosmology and Particle Physics Phenomenology suggest that while the CERN scientists did discover a unique new particle, there's no conclusive evidence of it being the Higgs boson."

"The Higgs could explain data obtained by CERN scientists using the Large Hadron Collider, but other particles could have created the data, suggesting there might be alternate explanations for it, they say in a paper published in the journal *Physical Review*: "The current data is not precise enough to determine exactly what the particle is," says university researcher Mads Toudal Frandsen, "It could be a number of other known particles."

What is CERN Really Doing?

CERN is the Home of the World Wide Web—Europe's PRISM

In 1989, under the guidance of Tim Berners-Lee, CERN began the World Wide Web project, which led to the first webpage in history. On April 30, 1993, CERN announced that the World Wide Web would be free to anyone. Essentially, CERN wrote the base code for individuals to access the Internet.

The first packet-switched computer network was the ARPANET. It used a backbone of routers called Interface Message Processors. Other packet-switched computer networks began to proliferate in the 1970s, eventually adopting TCP/IP protocols or being replaced by newer networks. The National Science Foundation created NSFNET in 1986 by funding six networking sites using links and peering at the ARPANET. In 1987, this new network was upgraded with links to thirteen sites.

These sites included regional networks that, in turn, connected over 170 other networks. IBM, MCI, and Merit upgraded the backbone bandwidth (T3) in 1991. The combination of the ARPANET and NSFNET became known as the Internet. Within a few years, the dominance of the NSFNet backbone led to the decommissioning of the redundant ARPANET infrastructure in 1990.

In the early days of the Internet, backbone providers exchanged their traffic at government-sponsored Network Access Points (NAPs), until the government privatized the Internet and transferred the NAPs to commercial providers.

A Network Access Point (NAP) is a public network exchange facility where Internet service providers (ISPs) connect with one another in peering arrangements. The NAPs were a key component in the transition from the 1990s NSFNET era (when many networks were government sponsored, and commercial traffic was prohibited) to the commercial Internet providers of today.

The four Network Access Points (NAPs) were defined under the U.S. National Information Infrastructure (NII) document as transitional data communications facilities at which Network Service Providers (NSPs) would exchange traffic, in replacement of the publicly financed NSFNET Internet backbone. The original four NAPs are in Washington, D.C., New York, Chicago, and California. As a transitional strategy, NAPs were effective, giving commercial network operators a bridge from the Internet's beginnings as a government-funded academic experiment to the modern Internet of many private-sector competitors collaborating to form a network-of-networks, anchored around Internet Exchange Points.

Cisco Systems—Child of DARPA

Cisco Systems, Inc. is an American multinational technology company headquartered in San Jose, California, that designs, manufactures, and sells networking equipment. It is considered to be the largest networking company in the world. Cisco was one of the first to sell commercially successful routers supporting multiple network protocols; routers that quickly became vital to Internet service providers and by, 1998, gave Cisco *de facto* monopoly in this critical segment. Essentially, every Internet message originally went through a Cisco router. In 2000, Cisco became the most valuable company in the world, with a market capitalization of more than US $500 billion.

Cisco's Linksys E2700, E3500, E4500 devices have been reported to be remotely updated to a firmware version that forces users to register for a cloud service, allowing Cisco to monitor their network use and ultimately shut down the cloud service account and thus render the affected router unusable. Cisco has been found to have built in a backdoor to all of their routers that connects them to the NSA's PRISM program of surveillance.

According to the German magazine *Der Spiegel* the NSA has developed JETPLOW for gaining access to ASA Cisco series 5505, 5510, 5520, 5540, and 5550 and 500-series PIX Firewalls. The National Security Agency's Tailored Access Operations (TAO) unit and other NSA employees intercept servers, routers, and other network gear being shipped to organizations targeted for surveillance and install covert firmware onto them before they're delivered. These Trojan horse systems were described by an NSA manager as being "some of the most productive operations in TAO."

Initially, the Network Access Points were controlled and operated by Cisco Systems in conjunction with DARPA and the National Security Agency. CERN is Europe's oldest and largest Network Access Point used by factions of the U.S. Central Intelligence Agency (among others) for surveillance of foreign countries and their citizens.

CERN also coordinates the work of Onyx, the Swiss intelligence gathering system maintained by the Federal Intelligence Service of Switzerland. The mission of Onyx is to monitor both civil and military communications via telephone, fax, Internet traffic, and satellite communications. Onyx uses lists of keywords to filter the intercepted content for information of interest. Onyx coordinates the work of other collection centers like Zimmerwald, Heimenschwand, and Leuk.

CERN is a Network Access Point for Surveillance for the ECHELON Spy System

ECHELON is a surveillance program (signals intelligence / SIGINT collection and analysis network) operated on behalf of the five signatory nations (Five Eyes) to the UKUSA Security Agreement -Australia, Canada, New Zealand, United Kingdom, and the United States. The program was created in the late 1960s to monitor the military and diplomatic communications of the Soviet Union and its Eastern Bloc allies during the Cold War and was formally established in 1971. By the end of the 20th century, the system referred to as ECHELON had allegedly evolved beyond its military and diplomatic origins to also become "...a global system for the interception of private and commercial communications."

The *Guardian* British newspaper summarized the capabilities of the ECHELON system as follows: "A global network of electronic spy stations that can eavesdrop on telephones, faxes, and computers. It can even track bank accounts. This information is stored on ECHELON computers, which can keep millions of records on individuals. Officially, however, Echelon doesn't exist."

In 2000, James Woolsey, the former Director of the U.S. Central Intelligence Agency, confirmed that U.S. intelligence uses interception systems and keyword searches to monitor European businesses. Some of the systems used to accomplish these tasks at CERN include:

ADVISE: (Analysis, Dissemination, Visualization, Insight, and Semantic Enhancement) is a research and development program within the United States Department of Homeland Security (DHS) Threat and Vulnerability Testing and Assessment (TVTA) portfolio. It is developing a massive data mining system which collects and analyzes data on everyone in the United States and performs a "threat analysis" of them. The data can be everything from financial records, phone records, emails, blog entries, website searches, and any other electronic information that can be put into a computer system.

Boundless Informant: NSA system to analyze global electronic information.

BULLRUN: NSA program to preserve its ability to eavesdrop on encrypted communications.

Carnivore: FBI system to monitor email and electronic communications.

DCSNet: FBI surveillance system that can perform instant wiretaps on any telecommunications device.

Fairview: A mass surveillance program directed at foreign mobile phone users.

Financial Crimes Enforcement Network: A bureau of the Department of the Treasury that collects and analyzes financial transactions in order to combat financial crimes.

ICREACH: Surveillance that is shared with 23 government agencies, including the CIA, DEA, and FBI, to search illegally collected personal records.

Magic Lantern: A keystroke logging software deployed by the FBI in the form of an e-mail attachment. When activated, it acts as a trojan horse and allows the FBI to decrypt user communications.

Main Core: A personal and financial database storing information of millions of U.S. citizens believed to be threats to national security.

MAINWAY: NSA database containing metadata for hundreds of billions of telephone calls made through the four largest telephone carriers in the United States.

MUSCULAR: Overseas wiretapping of Google and Yahoo unencrypted internal networks by the NSA.

MYSTIC: NSA voice interception program.

PRISM: NSA electronic surveillance program which can target customers of participating corporations outside or inside the United States.

Sentry Eagle: Program to monitor and attack an adversary's cyberspace.

Special Collection Service (SCS): It employs covert listening device technologies to bug foreign embassies, communications centers, computer facilities, fiber-optic networks, and government installations.

Stellar Wind: The name of four surveillance programs in one.

Tailored Access Operations: Intelligence-gathering unit of the NSA.

Turbulence: NSA information-technology project. It includes offensive cyberwarfare capabilities, like injecting malware into remote computers.

US Intelligence Community (IC): A cooperative federation of 16 government agencies working together, but also separately, to gather intelligence and conduct espionage.

Utah Data Center: The Intelligence Community's $1.5 billion data storage center that is designed to store extremely large amounts of data. CERN is the European Data Center.

X-Keyscore: NSA system for searching and analyzing Internet data about foreign nationals.

Information Awareness Office: An office established to bring together several DARPA projects focused on applying surveillance and information technology to track and monitor terrorists and other threats to U.S. national security.

ThinThread: NSA program involving wiretapping and sophisticated analysis of the resulting data.

Trailblazer Project: NSA program intended to develop a capability to analyze data carried on communications networks including cell phone networks, and the Internet.

Miami and its CERN-like Network Access Point

Miami's Network Access Point, located in central Miami, is called "The Cube." It is the Network Access Point (NAP) of the Americas, one of the world's largest data centers, which redirects most of the digital information that comes from Latin America. About 90% of data traffic from Central and South America passes through the South Florida facility before continuing to its final destination.

The NAP is, in short, one of the Internet's brains—facilitating people's online activity. There are 18 data centers around the world which play an important role in global digital communications. Security measures are especially strict at the NAP, and it is rare to be granted access.

The Cube, unlike other data centers, rents its equipment and space to private and public customers, so that they can share information between them. To prevent any service interruptions, the walls have seven -inch- thick, steel-reinforced concrete exterior panels in a hidden-from-sight building that does not have windows. Does the secretive description of the Cube NAP sound familiar? It sounds like what CERN brags about and says it is a leader in—data mining, surveillance, and housing the largest set of Cisco routers in Europe.

CERN is clearly a collection and operation center for ECHELON and the CIA that supports the trans-nationalists' agenda.

CERN is a hub for the World Wide Web PRISM data mining.

This is the true agenda of CERN—the "discovery" and "creation" of the World Wide Web. CERN is hidden from sight and as secure as possible in a country that is notorious for keeping secrets. CERN has discovered nothing and is only good for collecting data and analyzing it. Such far-fetched discoveries as CERN's recent analysis that "classical music exists in the fifth dimension and humans cannot go there" is just one of the ridiculous "discoveries" that the CERN super computers have been working on. This is, of course, complete hogwash, any researcher could have concluded as much in far less time and with far less money.

Let's Review the False Flag of All Time

CERN is the justification for the continued "mad science" that is conducted by theorists who have wasted trillions on Big Bang nonsense and particle physics that lead to nuclear weapons.

CERN is run by Warlord Bankers from Geneva, which is the home of the New World Order through the United Nations and the IMF, BIS, WCC, etc.

CERN discoveries are fake and not supported by other colliders, NASA, NSA, and many other scientists worldwide.

CERN does not have 17 miles of tunnels because there is no evidence they have ever built, and the photographs of CERN technical equipment demonstrate that they are not connected to the accelerator. The equipment is not in sterile environments, and they supposedly use the laboratory space for large gatherings, dancing, photo-ops, and education.

CERN's data could easily be produced by computer simulations, and it is not repeatable at other facilities.

CERN is a hoax, using its own definition of an Internet hoax by claiming authority via association, large numbers of scientists supposedly involved, and a critical life-and-death issue at the heart of the hoax.

CERN has collected trillions of dollars from many countries and yet only appears to have spent billions or even millions. The budgets are shared out between five French cities, two Swiss cities, CERN itself, and many subsidiary organizations. This certainly begs the question: Where is all the money going?

CERN is one of the original Network Access Points because it set up the source code for establishing the World Wide Web and it is the largest NAP in Europe for surveillance and data collection for the Swiss and French governments, and ECHELON.

There are many systems of surveillance coordinated through CERN, including Onyx, ECHELON, PRISM, and other similar DARPA and In-Q-Tel programs.

CERN is economically supported to find discoveries that can be weaponized—bombs, nuclear energy, etc.

CERN claims to be the most dangerous machine scientists have ever built. There are no good reasons to continue splitting atoms with a quark cannon.

CERN is based upon the efficacy of the Big Bang Theory, which is incorrect. Thus, CERN's research is fraud. CERN cannot prove anything because its basic assumptions are false and inconclusive, based on consensus rather than authentic and reproducible scientific research.

CERN fraudulently "discovered" the Higgs-Boson particle. Many non-CERN scientists say that CERN has discovered nothing and that all particle physics descriptions of atoms are completely theoretical and that other theories seem more viable and provable. There are four other accepted theories that supplant the Big Bang Theory. Thus, other theories make CERN and all atom smashers obsolete and dinosaurs of science; essentially, a complete waste of time, money, and energy, much like the ghost cities built in China—just another way to hide or launder money or infuse the dwindling base of the global financial Ponzi schemes with much-needed cash.

CERN plans to build another atom smasher ten times bigger than the current one that is already the largest in the world. This time, it will be built completely in Switzerland so the warlord bankers can capitalize on an even bigger money-laundering Ponzi scheme.

Rudolf Steiner's Axioms of Astrophysics

The Stars once spoke to Man
It is World destiny that they are silent now.
To be aware of the silence
Can become pain for earthly Man.
But in the deepening silence
There grows and ripens
What Man speaks to the Stars
To be aware of the speaking
Can be strength for Spirit-Man.

The science of astronomy is that science which, sooner than all others, may be restored to spirituality.

The heart of a person, however, is in effect a Universe turned inside out, and that is how the physical world is joined to the spiritual.

The universe is inwardly light; and observed from without, thought.

In light there is a fourth; it is of fourfold extension—it has the quality of inwardness as a fourth dimension.

The nature of light is thought, and that of darkness is will.

What lives around us in nature is the outcome of former moral occurrences, in the form of dying thought worlds manifesting as light.

Everything that happens in the universe is reflected by the Moon as if in a mirror.

The human body is indeed an image of the world of stars.

Man, as he stands on Earth is indeed a shrunken Universe.

The super-sensible form of man, which is created first in the ethereal and astral realm and only then shrinks and contracts till it is able to be clothed in physical material.

What is today the Sun was first a planet; and just as the Old Saturn became a Sun, so our Earth, now a planet, will gradually develop into a sun.

The formation of a cosmic body begins in spirit. Every physical cosmic body was first spirit.

The farther the Earth moves toward its sun-state, the more will this maya be consumed and perish in the fire of the Sun.

The human body is the most complex thing that exists anywhere in the universe.

All activities and influences are subject to the law of elasticity. The elasticity in them always has its boundary or limit.

Space and time, however, are not real things in themselves.

The world of fixed stars is, from our earthly perspective, the physical image or reflection of the spiritual world in which we live between death and a new birth, and which we then experience as our inner world.

A fixed star has not, from the very beginning, been a fixed star, but has had to pass through the lower school of planetary existence.

During the stage of Earth-existence itself, the Earth will reunite with the Sun, just as during the same phase of evolution it separated from the Sun.

The air-fire [warmth-light ether] beings sit fast, so to speak, in their airy strongholds and display their power and might in the fiery flashes of lightning [plasma].

Saturn gives the senses, Jupiter gives their continuation in the nerves, and Mars exerts the kind of control that endows man, for example, with the faculty of speech.

Whatever manifests as forces in that world comes from the Mars beings, whatever manifests as natural law [thought] from the Jupiter beings, and whatever manifests as color and sound from the Saturn beings.

Man can find absolutely nothing in the Universe unless he finds it first in himself.

The ability to think of the Universe as an organism depends on our learning to understand the hieroglyph of the organism we have before us. We must learn to perceive Man as a hieroglyph of the Universe. We must learn this first in Man, and we shall then find it in the Universe.

The living co-operation of the solar system produces in the center a hollow, which reflects as a mirror. That is the Sun!

The Sun is a hollow space of suction which annihilates everything within it. A space indeed that is less than hollow. A hollow space merely receives what is put into it; but the Sun is a hollow space of such a nature that anything brought to it is immediately absorbed and disappears. There in the Sun is not only nothing, but less than nothing. What shines to us in the light is the reflection of what first comes in from Cosmic space...

We have therefore not only to consider a world which is spread out before us, but another world which projects into, and permeates our own.

The Sun is not a ball of gas; but in that place where the Sun is, there is something less than empty space—a sucking, absorbing body, in fact, while all around it is that which exerts pressure. Consequently, in that which comes to us from the Sun we have not to do with anything constituting a product of combustion in the Sun; but all that has been transmitted to the Sun from the Universe is rayed back.

In the same way, not only does matter become empty space, but it becomes negative, less than nothing—emptier than emptiness; it assumes a 'sucking' nature. Ether is sucking, absorbing. Matter presses. Ether absorbs. The Sun is an absorbing, sucking ball, and wherever ether is present we have this absorbent force.

It is the undulating wave of pressure and suction, ponderable matter and ether, that forms in its alternation a living organization.

This revolution cannot take place in a manner that would allow us to look on the Sun's center as the pivot, but the pivot must be a neutral point situated between the center of the Sun and the center of the Earth.

There simply does not exist such a thing that mathematics calls 'Space'; but everywhere are lines of force, directions of force, and these are not equal, they vary, they are differentiated. He possesses them through his astral body, which establishes these relations. They are only possible to him because through his astral body man is a denizen of an astral world, a world which though acting upon space is not itself spatial. We only conceive the Zodiac in its full meaning when we treat it as the representative of the astral world beyond.

With all that is in him through his astral body and Ego, man is related to forces that are active beyond the Earth.

When we look at the Zodiac, it forms for us the outer cosmic representative of our own outer form.

That is, Earth and Sun move together in the course of the year; the one does not circle around the other.

The Earth moves behind the Sun in a screw-like line, the Earth moving along always with the Sun. And then, if we view the line from above, we get a projection of the line, and the projection shows a lemniscate.

He who wishes to study the human embryo from its first stage onwards must consider it as an image of the Cosmos.

The minutest molecule is acted upon by the whole starry heavens; the molecule is nothing but a focus of macrocosmic forces.

Empty space can be less than empty by being filled with ether, when it becomes a negative entity.

To understand the solar system, we must pass from the three dimensions into other dimensions.

The Sun's movement is an expression of a slower movement which meets a quicker movement, and we are so inserted into the two movements that the silhouette of our heart arises; and on to it the rest of the human form is fitted.

We look up to the planet saying: It has a definite form, that corresponds to the Spirits of Form; it has an inner mobility, that corresponds to the Spirits of Motion. All this is permeated by consciousness, which corresponds to the Spirits of Wisdom.

That which leads the planet through space, which governs its movement through space and causes it to revolve around the fixed star, corresponds to the Spirits of Willpower, or Thrones. The harmony of movement between one planet and another corresponds to the activity of the Cherubim.

In the First Hierarchy, which consists of the Thrones, Cherubim, and Seraphim, we have a form of creation in which the part created is detached—we have there not only self-creation, but world-creation. Creation of worlds is their external life—creation of beings their inner life.

Everything too which takes place in the kingdom of nature rhythmically and in recapitulation, such as day and night, alternations of the year, the four seasons of the year, everything which thus depends upon repeated happenings, is regulated by the Spirits of the Rotation of Time, the offspring of the beings of the First Hierarchy.

All physical matter comes into being through the meeting together of forces coming from the Spirits of Form. In this case, we have the meeting of the forces of the normal Spirits of Form and those of the abnormal Spirits of Form. They collide with one another and, in reality, an indentation is produced and consequently at this point a simultaneous breaking up of the form, but only of the form. The form breaks up and this hollow space is bored. Now, broken spiritual form, crushed form, is in reality matter.

With the light which flows in from the Sun, a notch is formed, and the planet throws itself against that light flowing in from the Sun; arrests it and throws it back.

Everything of a cometary or meteoric nature is ever striving to collect round itself all the harmful astral products, and to remove them out of the planetary system.

We must regard cometary life as something which continually works as a thunderstorm in the planetary system, cleansing it.

Uranus and Neptune did not originally belong to our Solar System; they came much later into the sphere of attraction of our system: they then joined company and remained within it.

The group-souls of the minerals are to be found in the sphere of the Thrones, those of the plants in the sphere of the Spirits of Wisdom, and the animals in the sphere of the Spirits of Motion; but man has so received his group-soul that with the inflowing of his ego, a group-soul was originally given him, as an emanation from the Spirits of Form.

He alone is studying reality, who on the one hand studies the starry skies and on the other hand the development of the human embryo.

Lunar life is just as much connected with the rhythmic system of man as is the solar life with his nerves-and-senses system.

What is alive in the planetary system is precisely what we cannot calculate.

Conceive of a space that is inherently mobile—or algebraic equations whose very functionality is in itself a function—and you will find the transition to a mathematical understanding of organic Nature.

That which is commonly thought of as the physical constitution of the Sun can never really be understood with the ideas we derive from earthly life.

Isis Sophia
Wisdom of God,
Lucifer has slain her
And upon the wings of the forces of the world
Has carried her away to distant realms.

The will of Christ
Working in the human being,
Will rescue from Lucifer
And on the boats of spiritual knowledge
Awaken in human souls
Isis-Sophia,
The wisdom of God.

Cosmological Charts

Diagram 1
Elohim †
Incarnations of the Earth

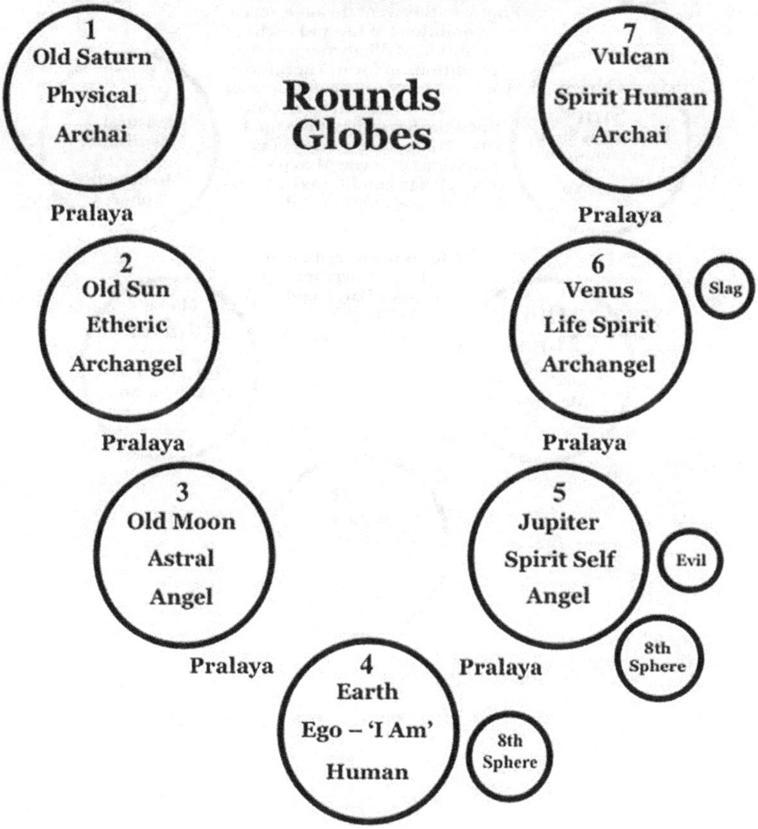

† Spirits of Form, Greek: Exousíai, ἐξουσίαι; Latin: Potestates; English: Powers, Authorities; Hebrew: ʾĔlōhīm, אֱלֹהִים - plural of ,אֱלוֹהַּ ʾĔlōah. "God(s), Heavenly Power(s)."

Diagram 2

7 Conditions of Consciousness

"Neither can they die any more: for they are equal unto the angels; and are the children of God, being the children of the resurrection." Luke (20:36)

"Each condition of consciousness can only run its course in seven conditions of life; each condition of life in seven conditions of form. That is 7 x 7 x 7 conditions. In fact, an entire evolution such as that of the Earth passes through 7 x 7 x 7 conditions of form. Our Earth was once Saturn; this went through seven conditions of life and each condition of life through seven conditions of form, Therefore you have forty-nine conditions of form upon Saturn, forty-nine upon the Sun, forty-nine upon the Moon, etc.; 7 x 49 = 343 conditions of form. Man passes through 343 conditions of form in the course of his evolution." †

"Jesus answered them, Is it not written in your law, I said, Ye are gods?" John (10:34)

† **Rudolf Steiner,**
***The Apocalypse of St. John*,**
Lecture X June 28th, 1908 (CW 104)

Diagram 3

7 Main Epochs of Earth Evolution

Polarian Epoch
Hyperborean Epoch
Lemurian Epoch
Atlantean Epoch
5th Post-Atlantean Epoch
6th Post-Atlantean Epoch
7th Post-Atlantean Epoch

7 Post-Atlantean Periods

1st Post-Atlantean Period
Old Indian Period
2nd Post-Atlantean Period
Old Persian Period
3rd Post-Atlantean Period
Egypto-Chaldean Period
4th Post-Atlantean Period
Greco-Roman Period
5th Post-Atlantean Period
Anglo-Germanic Period
6th Post-Atlantean Period
Russian Period
7th Post-Atlantean Period
American Period

The 5th Post-Atlantean Epoch is divided into 7 Post-Atlantean Periods of 2,160 years

Diagram 4

7 Atlantean Periods (Sub-Races)

1st Rmoahals
2nd Tlavtlis
3rd Toltecs
4th Original Turanians
5th Ur Semites
6th Akkadians
7th Mongolians

7 Post-Atlantean Periods

1st Old Indian Period
1st Old Persian Period
3rd Egypto-Chaldean-Babylonian-Hebrew Period
4th Greco-Roman Period
5th Anglo-Germanic Period
6th Russian Period
7th American Period

"Today the idea of civilization has already superseded the idea of race"
Rudolf Steiner, Munich, May 8th, 1907

"To this [Atlantean] Root-Race belong the following Sub-Races: Firstly the Rmoahals, secondly the Tlavtlis, thirdly the Toltecs, fourthly the original Turanians, fifthly the original Semites, sixthly the Akkadians, seventhly the Mongols. Still further back we come to the continent of Lemuria, between Africa, Asia and Australia." Rudolf Steiner, *The Foundations of Esotericism* Lecture XXIV – Berlin, October 26th, 1905 (CW 93a)

Diagram 5

7 Post-Atlantean Periods

The Age of Cancer **Old Indian Period**	♋	**The Etheric Body** **7,227 B.C.-5,067 B.C.**
The Age of Gemini **Old Persian Period**	♊	**The Astral Body** **5,067 B.C.-2,907 B.C.**
The Age of Taurus **Egypto-Chaldean Period**	♉	**Sentient-Soul** **2,907 B.C.-747 B.C.**
The Age of Aries **Greco-Roman Period**	♈	**Intellectual-Soul** **747 B.C.-1,414 A.D.**
The Age of Pisces **Anglo-Germanic Period**	♓	**Consciousness-Soul** **1,414 A.D.-3,574 A.D.**
The Age of Aquarius **E. European Period**	♒	**Spirit-Self** **3,574 A.D.-5,734 A.D.**
The Age of Capricorn **American Period**	♑	**Life-Spirit** **5,734 A.D.-7,894 A.D.**

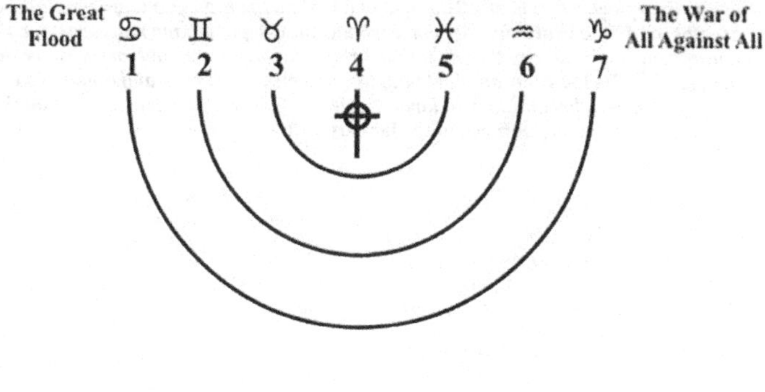

Diagram 6

The Cosmic Year – Post-Atlantean Epochs

5th Post Atlantean Period 7,227 B.C.-7,894 A.D.

1. ♋ 2. ♊ 3. ♉ 4. ♈ 5. ♓ 6. ♒ 7. ♑

6th Post Atlantean Period 7,894 A.D.-23,014 A.D.

1. ♐ 2. ♏ 3. ♎ 4. ♍ 5. ♌ 6. ♋ 7. ♊

7th Post Atlantean Period 23,014 A.D.-38,134 A.D.

1. ♉ 2. ♈ 3. ♓ 4. ♒ 5. ♑ 6. ♐ 7. ♏

"What is important for us is the fact that the position of the Sun at the spring equinox moves backward, passing through the whole Zodiac little by little. It traverses the whole Zodiac until it finally returns to the original position, taking approximately 25,920 years. These 25,920 years are termed the Platonic Year, the Cosmic Year. The exact figure varies according to the various methods of calculation. However, we are not concerned with exact figures but with the rhythm this precession entails. You can imagine that a cosmic rhythm must lie in this movement which repeats itself every 25,920 years. We can say that these 25,920 years are very important for the life of the Sun, for during this time the life of the Sun passes through one unit, a proper unit. The next 25,920 years are then a repetition. We have a rhythm in which one unit measures 25,920 years....

"On average, a human being takes eighteen breaths a minute. Not all breaths are equal, for those in youth differ from those in old age, but the average is eighteen breaths a minute. Eighteen times a minute we rhythmically renew our life. Multiply this by 60 and you have 1,080 times an hour. Now multiply by 24, and the number of breaths in twenty-four hours comes to 25,920!"

"A year has 365¼ days and if we divide 25,920 by 365.25 the answer is: nearly 71. Let us say 71 years, which is the average life-span of the human being. The human being is free, however, and often lives much longer; but you know that the patriarchal life-span is given as 70 years. The span of a human life is 25,920 days, 25,920 great breaths, and so we have another cycle wonderfully depicting the macrocosm in the microcosm. We could say that by living for one day, taking 25,920 breaths, we depict the Platonic Cosmic Year, and by living for 71 years, waking up and going to sleep 25,920 times — a breathing on a larger scale — we once again depict the Platonic Year." Rudolf Steiner, Dornach 28th, January 1916

Diagram 7

Nine-Fold Constitution of the Human Being

1. ***Old Saturn***	**Sphere of Planet Saturn**
Thrones – Warmth Ether	**Mineral Kingdom**
Spirits of Will	**4th Stage of the Archai**
Primal Physical Body as Warmth	
2. ***Old Sun***	**Sphere of Planet Jupiter**
Kyriotetes – Light Ether	**Plant Kingdom**
Spirits of Wisdom	**4th Stage of the Archangels**
Primal Etheric Body as Light	
3. ***Old Moon***	**Sphere of Planet Mars**
Dynamis – Sound Ether	**Animal Kingdom**
Spirits of Motion	**4th Stage of the Angels**
Primal Astral Body as Movement	
4. ***Earth***	**Human – Earth & Moon**
Exusiai – Life Ether	**Sentient-Soul**
Donate "I Am" – Ego	**Intellectual-Soul**
	Consciousness-Soul
5. ***Future Jupiter***	**Jupiter + Two Moons**
Heavenly Sound Ether	**Holy Spirit – Wisdom**
Angel Stage	**Spirit-Self (Manas)**
6. ***Future Venus***	**Venus + One Moon**
Heavenly Light Ether	**Son – Love**
Archangel Stage	**Life-Spirit (Budhi)**
7. ***Future Vulcan***	**Vulcan – No Moon**
Heavenly Warmth Ether	**Father – Will**
Archai Stage	**Spirit-Human (Atman)**

Recommended Reading

Gabriel, Tyla. *The Gospel of Sophia: The Biographies of the Divine Feminine Trinity,* Volume 1, Our Spirit, Northville, 2014.

Gabriel, Tyla. *The Gospel of Sophia: A Modern Path of Initiation,* Volume 2. Our Spirit, Northville, 2015.

Gabriel, Tyla and Douglas. *The Gospel of Sophia: Sophia Christos Initiation,* Volume 3. Our Spirit, Northville, 2016.

Gabriel, Douglas. *The Spirit of Childhood.* Trinosophia Press, Berkley, 1993.

Gabriel, Douglas. *Goddess Meditations.* Trinosophia Press, Berkley, 1994.

Gabriel, Douglas. *The Eternal Curriculum for Wisdom Children: Intuitive Learning and the Etheric Body.* Our Spirit, Northville, 2017.

Gabriel, Douglas. *The Eternal Ethers: A Theory of Everything.* Our Spirit, Northville, 2018.

Gabriel, Douglas. *Hidden History of the Grail Queens.* Our Spirit, Northville, 2019.

Gabriel, Douglas. *The Spirit of Childhood*: The Waldorf Curriculum. Our Spirit, Northville, 2024.

Gabriel, Douglas. *Intuitive Learning: Activating the Well-springs of Giftedness.* Our Spirit, Northville, 2023.

The Series: *From the Works of Dr. Rudolf Steiner*

Gabriel, Douglas. *The Incarnation of Ahriman: The Occult Annihilation of the Soul.* Our Spirit, Northville, 2023.

Gabriel, Douglas. *How to Become An Angel: Preparing for the Future Sixth and Seventh Epochs.* Our Spirit, Northville, 2023.

Gabriel, Douglas. *Surviving Apocalypse: Creating Your Ark.* Our Spirit, Northville, 2023.

Gabriel, Douglas. *Life After Death: Steiner's Book of the Dead.* Our Spirit, Northville, 2023.

Gabriel, Douglas. *The Second Coming of Christ in the Etheric Realm.* Our Spirit, Northville, 2024.

Gabriel, Douglas. *Grappling with Evil: Conquering Fear and Anxiety.* Our Spirit, Northville, 2024.

Gabriel, Douglas. *The Human Heart: Supersensible Organ of Perception.* Our Spirit, Northville, 2024.

Gabriel, Douglas. *Anthroposophia: The Divine Feminine Trinity.* Our Spirit, Northville, 2024.

Gabriel, Douglas. *Spiritual Enlightenment and Initiation.* Our Spirit, Northville, 2024.

Gabriel, Douglas. *Cosmic Intelligence and Spiritual Hierarchies.* Our Spirit, Northville, 2024.

Gabriel, Douglas. *Developing Christ in Our Threefold Ego.* Our Spirit, Northville, 2024.

Gabriel, Douglas. *The Eighth Sphere: Evil and Evolution.* Our Spirit, Northville, 2024.

Gabriel, Douglas. *Meditation for Enlightenment and Ascension.* Our Spirit, Northville, 2024.

Gabriel, Douglas. *Dweller on the Threshold: Encountering the Lord of Karma.* Our Spirit, Northville, 2024.

Gabriel, Douglas. *Rudolf Steiner and the Masters.* Our Spirit, Northville, 2024.

Gabriel, Douglas. *The Inklings and Star Wars.* Our Spirit, Northville, 2024.

Gabriel, Douglas. *Perception as Revelation of the Spirit.* Our Spirit, Northville, 2025.

Gabriel, Douglas. *The Etherization of the Blood.* Our Spirit, Northville, 2025.

Gabriel, Douglas. *The Perfected Bodies of Jesus Christ in the Realm of Spiritual Economy.* Our Spirit, Northville, 2025.

Gabriel, Douglas. *The Mystery of Life Found in Plasma.* Our Spirit, Northville, 2025.

Gabriel, Douglas. *The Resurrection Body of Christ and the Holy Grail.* Our Spirit, Northville, 2025.

Bibliography

Andreæ, Johann Valentin. *Reipublicæ Christianopolitanæ descriptio.* Argentorati: Sumptibus hæredum Lazari Zetzneri, Strasbourg, 1619.

Andreæ, Johann Valentin. *Johann Valentin Andreae's Christianopolis; an ideal state of the seventeenth century.* translated from the Latin of Johann Valentin Andreae with an historical introduction. by Felix Emil Held. The Graduate School of the University of Illinois, Urbana- Champaign, 1916.

Arnold, Edwin Sir. *The Light of Asia, or The Great Renunciation (Mahâbhinishkramana): Being the Life and Teaching Gautama, Prince of India and Founder of Buddhism (As Told in Verse by an Indian Buddhist).* Kegan Paul, Trench, Trübner & Co., London, 1879.

Avari, Burjor. *India: The Ancient Past: A History of the Indian Sub-Continent.* Routledge, New Edition, 2007.

Barnwell, John. *The Arcana of the Grail Angel: The Spiritual Science of the Holy Blood and of the Holy Grail.* Verticordia Press, Bloomfield Hills, 1999.

Barnwell, John. *The Arcana of Light on the Path: The Star Wisdom of the Tarot and Light on the Path.* Verticordia Press, Bloomfield Hills, 1999.

Blavatsky, H. P. (Helena Petrovna). *Isis Unveiled: A Master-Key to the Mysteries of Ancient and Modern Science and Theology*. J. W. Bouton. New York, 1878.

Blavatsky, H. P. (Helena Petrovna). *The Key to Theosophy: Being a Clear Exposition, in the Form of Question and Answer, of the Ethics, Science and Philosophy for the Study of Which the Theosophical Society Has Been Founded.* The Theosophical Publishing Company, Ltd. London, 1889.

Blavatsky, H. P. (Helena Petrovna). *The Secret Doctrine: The Synthesis of Science, Religion and Philosophy*. The Theosophical Publishing Company, Ltd. London, 1888.

Blavatsky, H. P. (Helena Petrovna). *The Voice of Silence: Being Extracts from the Book of the Golden Precepts*. Theosophical University Press, 1992.

Bockemuhl, Jochen. *Toward a Phenomenology of the Etheric World: Investigations into the Life of Nature and Man*. Anthroposophic Press, Spring Valley, N. Y., 1977.

Campanella, Tommaso. *The City of the Sun*. The ProjectGutenberg Ebook, David Widger, 2013.

Colum, Padriac. *Orpheus: Myths of the World*. Floris Books. Colum, Padriac. The Children's Homer. MacMillan Co., 1946.

Colum, Padriac. *The Tales of Ancient Egypt*. Henry Walck Incorporated, New York,1968.

Crawford, John Martin. *The Kalevala: The Epic Poem of Finland*. John B. Alden, New York, 1888.

Gabriel, Douglas. *The Eternal Curriculum for Wisdom Children: Intuitive Learning and the Etheric Body*. Our Spirit, Northville, 2017.

Gabriel, Tyla. *The Gospel of Sophia: The Biographies of the Divine Feminine Trinity*, Volume 1, Our Spirit, Northville, 2014.

Gabriel, Tyla. *The Gospel of Sophia: A Modern Path of Initiation,* Volume 2. Our Spirit, Northville, 2015.

Gabriel, Tyla and Douglas. *The Gospel of Sophia: Sophia Christos Initiation,* Volume 3. Our Spirit, Northville, 2016.

Gabriel, Douglas. *The Spirit of Childhood.* Trinosophia Press, Berkley, 1993.

Gabriel, Douglas. *The Eternal Ethers: A Theory of Everything.* Our Spirit, Northville, 2018.

Gabriel, Douglas. *Goddess Meditations.* Trinosophia Press, Berkley, 1994.

Gebser, Jean. *The Ever Present Origin.* Ohio University Press, 1991.

Green, Roger Lancelyn & Heather Copley. *Tales of Ancient Egypt.* Puffin Books, New York, 1980.

Harrison, C. G. *The Transcendental Universe; Six Lectures on Occult Science, Theosophy, and the Catholic Faith.* George Redway, London 1893.

Harrison, C. G. *The Transcendental Universe; Six Lectures on Occult Science, Theosophy, and the Catholic Faith.* Delivered Before the Berean Society, edited with an introduction by Christopher Bamford. Lindesfarne Press, Hudson, 1993.

Hamilton, Edith. *Mythology.* Little Brown And Co., Boston, 1942.

Harrer, Dorothy. *Chapters from Ancient History.* Waldorf Publications, Chatham, 2016.

Hazeltine, Alice Isabel. *Hero Tales from Many Lands.* Abingdon Press, New York, 1961.

Heidel, Alexander. *The Babylonian Genesis: The Story of Creation.* University of Chicago Press, Chicago, 1942.

Hiebel, Frederick. *The Gospel of Hellas.* Anthroposophic Press, New York, 1949.

Jocelyn, Beredene. *Citizens of the Cosmos: Life's Unfolding from Conception through Death to Rebirth.* Continuum, New York, 1981.

König, Karl. *Earth and Man.* Bio-Dynamic Literature, Wyoming, Rhode Island, 1982.

Kovacs, Charles. *Ancient Mythologies and History.* Resource Books, Scotland, 1991.

Kovacs, Charles. *Greek Mythology and History.* Resource Books, Scotland, 1991.

Landscheidt, Theodor. *Sun-Earth-Man a Mesh of Cosmic Oscillations: How Planets Regulate Solar Eruptions, Geomagnetic Storms, Conditions of Life, and Economic Cycles.* Urania Trust, London, 1989.

Laszlo, Ervin and Kingsley, Dennis L. *Dawn of the Akashic Age: New Consciousness, Quantum Resonance, and the Future of the World.* Inner Traditions, Rochester Vermont, 2013

Plato. *The Republic.* Dover Thrift Editions, 2000.

Sister Nivedita (Margaret E. Noble) & Coomaraswamy, Ananda K.. *Myths of the Hindus and Buddhists.* Henry Holt, New York 1914.

Steiner, Rudolf. *Ancient Myths: Their Meaning and Connection with Evolution.* Steiner Book Center, 1971.

Steiner, Rudolf. *Christ and the Spiritual World: The Search for the Holy Grail.* Rudolf Steiner Press, London, 1963.

Steiner, Rudolf. *Foundations of Esotericism.* Rudolf Steiner Press, London, 1983.

Steiner, Rudolf. *Isis Mary Sophia: Her Mission and Ours.* Steiner Books, 2003.

Steiner, Rudolf. *Man as a Being of Sense and Perception.* Steiner Book Center, Vancouver, 1981.

Steiner, Rudolf. *Man as Symphony of the Creative Word*. Rudolf Steiner Publishing, London, 1978.

Steiner, Rudolf. *Occult Science*. Anthroposophic Press, NY, 1972.

Steiner, Rudolf. *Rosicrucian Esotericism*. Anthroposophic Press, NY, 1978.

Steiner, Rudolf. *Rosicrucian Wisdom: An Introduction*. Rudolf Steiner Press, London, 2000. GA 425

Steiner, Rudolf. *The Bridge between Universal Spirituality and the Physical Constitution of Man*. Anthroposophic Press, NY, 1958.

Steiner, Rudolf. *The Evolution of Consciousness*. Rudolf Steiner Press, London, 1926.

Steiner, Rudolf. *The Goddess from Natura to the Divine Sophia*. Sophia Books, 2001.

Steiner, Rudolf. *The Holy Grail: from the Works of Rudolf Steiner.* Compiled by Steven Roboz. Steiner Book Center, North Vancouver, 1984.

Steiner, Rudolf. *The Influence of Spiritual Beings Upon Man*. Anthroposophic Press, NY, 1971.

Steiner, Rudolf. *The Reappearance of Christ in the Etheric*. Anthroposophic Press, NY, 1983.

Steiner, Rudolf. *The Risen Christ and the Etheric Christ*. Rudolf Steiner Press, London, 1969.

Steiner, Rudolf. *The Search for the New Isis the Divine Sophia*. Mercury Press, N.Y., 1983.

Steiner, Rudolf. *The Spiritual Hierarchies and the Physical World.* Anthroposophic Press, N.Y., 1996.

Steiner, Rudolf. *The Tree of Life and the Tree of Knowledge*. Mercury Press, NY, 2006.

Steiner, Rudolf. *The True Nature of the Second Coming*. Rudolf Steiner Press, London, 1971.

Steiner, Rudolf. *Theosophy*. Anthroposophic Press. New York, 1986.

Steiner, Rudolf. *Wonders of the World, Ordeals of the Soul, Revelations of the Spirit*. Rudolf Steiner Press, London, 1963.

Steiner, Rudolf. *World History in Light of Anthroposophy*. Rudolf Steiner Press, London, 1977.

Tappan, Eva March. *The Story of the Greek People*. Houghton Mifflin Co., Boston 1908.

van Bemmelen, D. J. *Zarathustra: The First Prophet of Christ*, 2 Vols. Uitgeverij Vrij Geestesleven, The Netherlands, 1968.

Watson, Jane Werner (Vālmīki). *Rama of the Golden Age: An Epic of India*. Garrard Pub., Champaign.

About Dr. Rudolf Steiner

Rudolf Steiner was born on the 27th of February 1861 in Kraljevec in the former Kingdom of Hungary and now Croatia. He studied at the College of Technology in Vienna and obtained his Doctorate at the University of Rostock with a dissertation on Theory of Knowledge which concluded with the sentence: "The most important problem of human thinking is this: to understand the human being as a free personality, whose very foundation is himself."

He exchanged views widely with the personalities involved in cultural life and arts of his time. However, unlike them, he experienced the spiritual realm as the other side of reality. He gained access through exploration of consciousness using the same method as the natural scientist uses in the visible world for his external research. This widened perspective enabled him to give significant impulses in many areas such as art, pedagogy, curative education, medicine, agriculture, architecture, economics, and social sciences, aiming towards the spiritual renewal of civilization.

He gave his movement the name of "Anthroposophy" (the wisdom of humanity) after separating from the German section of the Theosophical Society, where he had acted as a general secretary. He then founded the Anthroposophical Society in 1913 which formed its center with the construction of the First Goetheanum in Dornach, Switzerland. Rudolf Steiner died on 30th March 1925 in Dornach. His literary work is made up of numerous books, transcripts and approximately 6000 lectures which have for the most part been edited and published in the Complete Works Edition.

Steiner's basic books, which were previously a prerequisite to gaining access to his lectures, are: *Theosophy, The Philosophy of Freedom, How to Know Higher Worlds, Christianity as a Mystical Fact,* and *Occult Science.*

About the Author, Dr. Douglas Gabriel

Dr. Gabriel is a retired superintendent of schools and professor of education who has worked with schools and organizations throughout the world. He has authored many books ranging from teacher training manuals to philosophical/spiritual works on the nature of the divine feminine.

He was a Waldorf class teacher and administrator at the Detroit Waldorf School and taught courses at Mercy College, the University of Detroit, and Wayne State University for decades. He then became the Headmaster of a Waldorf School in Hawaii and taught at the University of Hawaii, Hilo. He was a leader in the development of charter schools in Michigan and helped found the first Waldorf School in the Detroit Public School system and the first charter Waldorf School in Michigan.

Gabriel received his first degree in religious formation at the same time as an associate degree in computer science in 1972. This odd mixture of technology and religion continued throughout his life. He was drafted into and served in the Army Security Agency (NSA) where he was a cryptologist and systems analyst in signal intelligence, earning him a degree in signal broadcasting. After military service, he entered the Catholic Church again as a Trappist monk and later as a Jesuit priest where he earned PhD's in philosophy and comparative religion, and a Doctor of Divinity. He came to Detroit and earned a BA in anthroposophical studies and history and a MA in school administration. Gabriel left the priesthood and became a Waldorf class teacher and administrator in Detroit and later in Hilo, Hawaii.

Douglas has been a sought-after lecturer and consultant to schools and businesses throughout the world and in 1982 he founded the Waldorf Educational Foundation that provides funding for the publication of educational books. He has raised a great deal of money for Waldorf schools and institutions that continue to develop the teachings of Dr. Rudolf Steiner. Douglas is now retired but continues to write a variety of books including a novel and a science fiction thriller. He has four children, who keep him busy and active and a wife who is always striving towards the spirit through creating an "art of life." She is the author of the Gospel of Sophia trilogy.

The Gabriels' articles, blogs, and videos can currently be found at: OurSpirit.com, Neoanthroposphy.com, GospelofSophia.com, EternalCurriculum.com

Made in the USA
Monee, IL
13 September 2026

62981883R00148